国际食品法典
乳及乳制品标准汇编

● 郑蔚然　杨　华　主译

中国农业科学技术出版社

图书在版编目（CIP）数据

国际食品法典乳及乳制品标准汇编 / 郑蔚然，杨华主译. --北京：中国农业科学技术出版社，2022.11
ISBN 978-7-5116-5544-8

Ⅰ.①国… Ⅱ.①郑…②杨… Ⅲ.①食品标准-汇编-世界 ②鲜乳-食品标准-汇编-世界 ③乳制品-食品标准-汇编-世界 Ⅳ.①TS207.2 ②TS252-65

中国版本图书馆CIP数据核字（2021）第211842号

责任编辑　王惟萍
责任校对　马广洋
责任印制　姜义伟　王思文

出 版 者	中国农业科学技术出版社
	北京市中关村南大街12号　邮编：100081
电　　话	（010）82106643（编辑室）　（010）82109702（发行部）
	（010）82109709（读者服务部）
网　　址	https://castp.caas.cn
经 销 者	各地新华书店
印 刷 者	北京中科印刷有限公司
开　　本	185 mm×260 mm　1/16
印　　张	23.25
字　　数	605千字
版　　次	2022年11月第1版　2022年11月第1次印刷
定　　价	89.80元

――― 版权所有・侵权必究 ―――

《国际食品法典乳及乳制品标准汇编》
译者名单

主　译：郑蔚然　杨　华
副主译：褚田芬　高　芳　于国光
译　者：（按姓氏笔画排序）
　　　　于国光　王小骊　戈惠可　吉小凤
　　　　朱汉鑫　任霞霞　刘　琳　刘玉红
　　　　李　辉　李勤锋　杨　华　肖英平
　　　　汪　雯　陈　敏　陈文明　金　诺
　　　　郑蔚然　胡　飖　钱文婧　高　芳
　　　　盛　慧　雷　玲　褚田芬　蔡　铮

国际食品法典委员会（Codex Alimentarius Commission，CAC）是由联合国粮食及农业组织（Food and Agriculture Organization，FAO）和世界卫生组织（World Health Organization，WHO）于1963年联合成立的政府间合作组织，旨在通过建立国际统一的农产品及食品安全和质量标准、加工规范和准则，提高国际食品贸易的安全性和公正性，保护消费者的身体健康和生命安全。目前，CAC已被世界贸易组织（World Trade Organization，WTO）认定为三个农产品及食品国际标准化机构之一。

《国际食品法典》汇集了国际食品法典委员会已批准的标准、规范准则和其他建议，规范对象包括食品原料、加工及半加工食品，具体内容涉及卫生和质量等，包括微生物指标、食品添加剂、农药与兽药残留、污染物、标签及产品说明、抽样和分析方法等。中国作为WTO成员，扮演着国际农产品及食品贸易大国的重要角色，日益深入地参与国际活动。基于此，我们认为将国际食品法典系统介绍给读者具有重要意义，因此，我们对国际食品法典委员会制定的乳及乳制品标准进行了系统性收集和整理，并编译成《国际食品法典乳及乳制品标准汇编》供我国读者参考使用。真诚希望该书的出版发行能对从事乳及乳制品生产、加工、检测、贸易、研究和标准制定的人士提供一定的帮助。

本书英文资料来源于国际食品法典官方网站，由于英文资料呈现方式具有一定差异性，我们在翻译时对中文翻译进行了统一。资料收集截止时间为2021年6月。由于译者水平和能力有限，本书在内容收集及文字翻译上可能存在疏漏和不当之处，敬请各位读者批评和指正。

译　者
2021年10月

CONTENTS

乳品术语 …………………………………………………………………………………… 1

GENERAL STANDARD FOR THE USE OF DAIRY TERMS ……………………… 4

乳粉和奶油粉 ……………………………………………………………………………… 7

STANDARD FOR MILK POWDERS AND CREAM POWDER ………………………… 12

盐水干酪 …………………………………………………………………………………… 17

GROUP STANDARD FOR CHEESES IN BRINE ………………………………………… 20

未成熟干酪（包括新鲜干酪）…………………………………………………………… 23

GROUP STANDARD FOR UNRIPENED CHEESE INCLUDING FRESH CHEESE……… 29

发酵乳制品 ………………………………………………………………………………… 35

STANDARD FOR FERMENTED MILKS ………………………………………………… 47

脱脂乳与植物脂肪混合物 ………………………………………………………………… 60

STANDARD FOR A BLEND OF EVAPORATED SKIMMED MILK AND VEGETABLE FAT ……………………………………………………………………………… 64

脱脂乳与植物脂肪混合粉剂 ……………………………………………………………… 68

STANDARD FOR A BLEND OF SKIMMED MILK AND VEGETABLE FAT IN POWDERED FORM ………………………………………………………………………… 73

脱脂甜炼乳与植物脂肪混合物 …………………………………………………………… 78

STANDARD FOR A BLEND OF SWEETENED CONDENSED SKIMMED MILK AND VEGETABLE FAT ……………………………………………………………………… 82

乳脂涂抹物 ………………………………………………………………………………… 87

STANDARD FOR DAIRY FAT SPREADS ………………………………………………… 94

马苏里拉干酪 ··· 102

STANDARD FOR MOZZARELLA ·· 110

切达干酪 ··· 119

STANDARD FOR CHEDDAR ·· 124

旦伯干酪 ··· 129

STANDARD FOR DANBO ·· 133

艾丹姆干酪 ··· 137

STANDARD FOR EDAM ·· 142

高达干酪 ··· 147

STANDARD FOR GOUDA ·· 152

哈瓦蒂干酪 ··· 157

STANDARD FOR HAVARTI ·· 161

萨姆索干酪 ··· 165

STANDARD FOR SAMSØ ·· 169

埃门塔尔干酪 ··· 173

STANDARD FOR EMMENTAL ·· 178

泰尔西特干酪 ··· 184

STANDARD FOR TILSITER ·· 188

圣宝林干酪 ··· 192

STANDARD FOR SAINT-PAULIN ··· 197

菠萝伏洛干酪 ··· 202

STANDARD FOR PROVOLONE ·· 207

农家干酪 ··· 212

STANDARD FOR COTTAGE CHEESE ··· 219

库洛米耶尔干酪 ··· 226

STANDARD FOR COULOMMIERS ··· 231

奶油干酪 ··· 236

STANDARD FOR CREAM CHEESE ·· 244

卡门培尔干酪 ··· 252

STANDARD FOR CAMEMBERT ··· 257

布里干酪	263
STANDARD FOR BRIE	268
特硬搓碎干酪	273
STANDARD FOR EXTRA HARD GRATING CHEESE	275
黄油	277
STANDARD FOR BUTTER	279
乳脂产品	281
STANDARD FOR MILKFAT PRODUCTS	284
炼乳	287
STANDARD FOR EVAPORATED MILKS	291
甜炼乳	295
STANDARD FOR SWEETENED CONDENSED MILKS	299
干酪通则	303
GENERAL STANDARD FOR CHEESE	309
乳清干酪	316
STANDARD FOR WHEY CHEESES	319
稀奶油和预制稀奶油	322
STANDARD FOR CREAM AND PREPARED CREAMS	330
乳清粉	338
STANDARD FOR WHEY POWDERS	341
食用酪蛋白制品	344
STANDARD FOR EDIBLE CASEIN PRODUCTS	349
乳制品渗透物粉	354
STANDARD FOR DAIRY PERMEATE POWDERS	357

乳品术语

GENERAL STANDARD FOR THE USE OF DAIRY TERMS

CXS 206-1999[1]

1999年通过。

1　范围

本标准适用于即食或用于进一步加工的乳品使用术语。

2　说明

2.1　乳是指哺乳动物的正常乳腺分泌物，通过一次或多次挤奶获得。即未添加且未从中提取任何成分，作为液态乳形式饮用或用于进一步加工的食品原料。

2.2　乳制品是指乳通过任何加工处理得到的产品。其可能含食品添加剂和加工所需的其他功能成分。

2.3　复合乳制品是指以乳、乳制品或乳组分为基本组成部分的产品，其中非乳组分不用于全部或部分替代为乳组分。

2.4　复原乳制品是指一种干燥或浓缩产品中添加适量水分，以重建其中水与固体比例所形成的产品。

2.5　再制乳制品是指乳脂和非脂乳固体混合而形成的产品，添加或未添加水分均可，以达到适当乳产品成分。

2.6　乳品术语是指直接、间接提到或暗示乳品或乳制品的名称、类别、符号、图案或其他装置的专业术语。

3　总则

应该正确使用乳品术语来描述或介绍食品，以避免误导消费者或使其困惑，并确保食品贸易中的公平交易。

4　乳品术语的应用

4.1　总要求

4.1.1　产品名称应按照《预包装食品标识通用标准》（CXS 1-1985）第4.1条规定进行申报。

4.1.2　来源于单种或多种动物乳汁的产品，应在产品名称前后插入指示相应的动物名称。若省略不会误导消费者，则无须声明。

---

[1] 本标准取代关于乳及乳制品的国际食品法典标准准则。

4.2 术语"乳"的应用

4.2.1 只有符合本标准第2.1条定义的食品才能被称为"乳"。若以此形式销售的食品，可以称为"生乳（或原料乳）"，或用不误导、混淆消费者的其他适当术语。

4.2.2 通过添加和/或去除乳组分对乳成分进行修改，并且对乳的这种修改有明确描述，放在其名称相近处，此产品可使用一个含有"乳"的术语命名。

4.2.3 尽管有本标准第4.2.2条的规定，但调整乳脂肪和/或蛋白质含量，并可即食的产品，也可称为"乳"，但要具备以下条件：

— 仅在允许做此种调整的产品销售国销售该产品；

— 调整后乳脂肪和/或蛋白质的最低和最高含量应符合产品销售国立法规定。在此情况下，其蛋白质含量应在该国规定范围内；

— 按照产品销售国立法允许的方法调整乳成分，仅通过添加和/或去除乳组分而不改变其乳清蛋白与酪蛋白的比例；

— 调整依据本标准第4.2.2条规定。

4.3 食品法典商品标准中乳制品术语的应用

4.3.1 只有产品符合国际食品法典委员会乳制品标准规定，才能按其指定名称命名。

4.3.2 尽管有本标准第4.3.1条和《预包装食品标识通用标准》（CXS 1-1985）第4.1.2条规定，当某个乳制品经加工后，其脂肪和/或蛋白质调整符合相关标准中乳制品成分的指标，也可用国际食品法典标准指定的乳制品名称命名。

4.3.3 通过添加和/或去除乳组分调整乳成分的产品，如果保留该产品基本特征，此成分修改在有关国家食品法典标准限度内，且使用量适当描述，可用有关乳制品名称命名。

4.4 复原乳及再制乳制品术语的应用

当乳和乳制品是由《预包装食品标识通用标准》（CXS 1-1985）第4.1.2条再制乳、复原乳或其乳制品制成，如果不会误导消费者或使消费者混淆，就可以用乳制品国际食品法典标准中规定的相关名称命名。

4.5 复合乳制品术语的应用

符合第2.3条描述的产品可用"乳"命名，如果在其名称附近对其他特殊配料（如调味食品、香料、香草和调味料）给予明确描述，也可用乳制品指定名称命名。

4.6 其他食品中乳制品术语的应用

4.6.1 在第4.2～4.5条提到的名称仅可在乳、乳制品或复合乳制品名称或其标识中使用。

4.6.2 第4.6.1条的规定不适用那些与传统习惯用法有不同特性的产品名称或该名称已明确地用于描述某种非乳制品特性的产品。

4.6.3 对于那些不属于乳、乳制品或复合乳制品的产品，在其标识、商业文件、宣传资料或在以任何形式的销售介绍中，不应命名或暗示该产品为乳、乳制品或复合乳制品[1]。

4.6.4 关于第4.6.3条提到的产品，如果产品中含乳、乳制品或乳组分，这些乳成分又是该产品的主要组分，且其来源于非乳成分并未用来部分或全部代替任何乳成分，则可以使用"乳"或乳制品名称来描述产品特性。仅在消费者不被误导的情况下，才可使用乳品术语。

如果是打算用其最终产品来代替乳、乳制品或复合乳制品，则不能使用乳品术语。

第4.6.3条提到的产品，如果就产品特征而言，其中含乳、乳制品或乳组分，这些乳组分又非该产品主要成分，则只能在配料表中按照《预包装食品标识通用标准》（CXS 1-1985）的规定使用乳品术语，而不能将乳品术语用于其他目的。

5 预包装食品标识

预包装乳、乳制品和复合乳制品的标识，除去某一特定国际食品法典标准或在本标准第4条有明确规定外，应按照《预包装食品标识通用标准》（CXS 1-1985）第4条规定进行标识。

---

[1] 在不误导消费者的情况下，《预包装食品标识通用标准》（GSLPF）第 4.1.13 条定义的描述名称和 GSLPF 第 4.2.1.2 条定义的配料名单不包括在内。

GENERAL STANDARD FOR THE USE OF DAIRY TERMS

CXS 206-1999[1]

Adopted in 1999.

1. **SCOPE**

   This General Standard applies to the use of dairy terms in relation to foods to be offered to the consumer or for further processing.

2. **DEFINITIONS**

2.1 **Milk** is the normal mammary secretion of milking animals obtained from one or more milkings without either addition to it or extraction from it, intended for consumption as liquid milk or for further processing.

2.2 **Milk product** is a product obtained by any processing of milk, which may contain food additives, and other ingredients functionally necessary for the processing.

2.3 **Composite milk product** is a product of which the milk, milk products or milk constituents are an essential part in terms of quantity in the final product, as consumed provided that the constituents not derived from milk are not intended to take the place in part or in whole of any milk constituent.

2.4 **A reconstituted milk product** is a product resulting from the addition of water to the dried or concentrated form of the product in the amount necessary to re-establish the appropriate water to solids ratio.

2.5 **A recombined milk product** is a product resulting from the combining of milkfat and milk-solids-non-fat in their preserved forms with or without the addition of water to achieve the appropriate milk product composition.

2.6 **Dairy terms** means names, designations, symbols, pictorial or other devices which refer to or are suggestive, directly or indirectly, of milk or milk products.

3. **GENERAL PRINCIPLES**

   Foods shall be described or presented in such a manner as to ensure the correct use of dairy terms intended for milk and milk products, to protect consumers from being confused or misled and to ensure fair practices in the food trade.

4. **APPLICATION OF DAIRY TERMS**

4.1 **General requirements**

4.1.1 The name of the food shall be declared in accordance with Section 4.1 of the *General Standard for the Labelling of Prepackaged Foods* (CXS 1-1985).

4.1.2 A word or words denoting the animal or, in the case of mixtures, all animals from which the milk has been derived shall be inserted immediately before or after the designation of the product. Such declarations are not required if the consumer would not be misled

---

[1] This Standard replaced the Code of Principles Concerning Milk and Milk Products.

by their omission.

4.2 Use of the term milk

4.2.1 Only a food complying with the definition in Section 2.1 may be named "milk". If such a food is offered for sale as such it shall be named "raw milk" or other such appropriate term as would not mislead or confuse the consumer.

4.2.2 Milk which is modified in composition by the addition and/or withdrawal of milk constituents may be identified with a name using the term "milk", provided that a clear description of the modification to which the milk has been subjected is given in close proximity to the name.

4.2.3 Notwithstanding the provisions of Section 4.2.2 of this Standard, milk which is adjusted for fat and/or protein content and which is intended for direct consumption, may also be named "milk" provided that:

- it is sold only where such adjustment is permitted in the country of retail sale;
- the minimum and maximum limits of fat and/or protein content (as the case may be) of the adjusted milk are specified in the legislation of the country of retail sale. In this case the protein content shall be within the limits of natural variation within that country;
- the adjustment has been performed according to methods permitted by the legislation of the country of retail sale, and only by the addition and/or withdrawal of milk constituents, without altering the whey protein to casein ratio; and
- the adjustment is declared in accordance with Section 4.2.2 of this standard.

4.3 Use of the names of milk products in Codex commodity standards

4.3.1 Only a product complying with the provisions in a Codex standard for a milk product may be named as specified in the Codex standard for the product concerned.

4.3.2 Notwithstanding the provisions of Section 4.3.1 of this Standard and Section 4.1.2 of the *General Standard for the Labelling of Prepackaged Foods* (CXS 1-1985), a milk product may be named as specified in the Codex standard for the relevant milk product when manufactured from milk, the fat and/or protein content of which has been adjusted, provided that the compositional criteria in the relevant standard are met.

4.3.3 Products that are modified through the addition and/or withdrawal of milk constituents may be named with the name of the relevant milk product in association with a clear description of the modification to which the milk product has been subjected provided that the essential product characteristics are maintained and that the limits of such compositional modifications shall be detailed in the standards concerned as appropriate.

4.4 Use of terms for reconstituted and recombined milk products

Milk and milk products may be named as specified in the Codex Standard for the relevant milk product when made from recombined or reconstituted milk or from recombination or reconstitution of milk products in accordance with Section 4.1.2 of the *General Standard for the Labelling of Prepackaged Foods* (CXS 1-1985), if the consumer would not be misled or confused.

4.5　**Use of terms for composite milk products**

A product complying with the description in Section 2.3 may be named with the term "milk" or the name specified for a milk product as appropriate, provided that a clear description of the other characterizing ingredient(s) (such as flavouring foods, spices, herbs and flavours) is given in close proximity to the name.

4.6　**Use of dairy terms for other foods**

4.6.1　The names referred to in Sections 4.2 to 4.5 may only be used as names or in the labelling of milk, milk products or composite milk products.

4.6.2　However, the provision in Section 4.6.1 shall not apply to the name of a product the exact nature of which is clear from traditional usage or when the name is clearly used to describe a characteristic quality of the non-milk product.

4.6.3　In respect of a product which is not milk, a milk product or a composite milk product, no label, commercial document, publicity material or any form of point of sale presentation shall be used which claims, implies or suggests that the product is milk, a milk product or a composite milk product, or which refers to one or more of these products[1].

4.6.4　However, with regard to products referred to in Section 4.6.3, which contain milk or a milk product, or milk constituents, which are an essential part in terms of characterization of the product, the term "milk", or the name of a milk product may be used in the description of the true nature of the product, provided that the constituents not derived from milk are not intended to take the place, in part or in whole, of any milk constituent. For these products dairy terms may be used only if the consumer would not be misled.

If however the final product is intended to substitute milk, a milk product or composite milk product, dairy terms shall not be used.

For products referred to in Section 4.6.3 which contain milk, or a milk product, or milk constituents, which are not an essential part in terms of characterization of the product, dairy terms can only be used in the list of ingredients, in accordance with the *General Standard for the Labelling of Prepackaged Foods* (CXS 1-1985). For these products dairy terms cannot be used for other purposes.

5　**LABELLING OF PREPACKAGED FOODS**

Prepackaged milk, milk products and composite milk products shall be labelled in accordance with Section 4 of the *General Standard for the Labelling of Prepackaged Foods* (CXS 1-1985), except to the extent otherwise expressly provided in a specific Codex standard or in Section 4 of this Standard.

1　This excludes descriptive names as defined in Section 4.1.1.3 of the *General Standard for the Labelling of Prepackaged Foods* (GSLPF) and ingredients lists as defined in Section 4.2.1.2 of the GSLPF providing the consumer would not be misled.

乳粉和奶油粉

STANDARD FOR MILK POWDERS AND CREAM POWDER

CXS 207-1999

本标准取代《全脂乳粉、部分脱脂乳粉和脱脂乳粉标准》（A-5-1971）和《奶油粉、半奶油粉和高脂乳粉标准》（A-10-1971）。

1999年通过。2010年、2013年、2014年、2016年、2018年修订。

1 范围

本标准适用于符合本标准第2条所述即食或用于进一步加工的乳粉和奶油粉。

2 说明

乳粉和奶油粉是指将牛乳或稀奶油去除部分水分后得到的产品。为符合本标准第3条对成分的要求，可通过添加和/或去除某些奶组分以调整奶或奶油中的脂肪和/或蛋白质含量，而其乳清蛋白与酪蛋白的比例不变。

3 基本成分和质量要求

3.1 基本原料

牛乳和奶油。

允许用下列乳制品调整蛋白质：

- 乳超滤滞留物：乳超滤滞留物是指乳、部分脱脂乳或脱脂乳，通过超滤法浓缩乳蛋白后形成的产品；
- 乳渗透物：乳渗透物是指通过超滤法，将乳蛋白和乳脂从乳、部分脱脂乳或脱脂乳中分离除去而形成的产品；
- 乳糖[1]。

3.2 成分

奶油粉

乳脂最低含量	42% m/m
水分最高含量[a]	5% m/m
乳的非脂固体中乳蛋白最低含量[a]	34% m/m

---

1 参见《糖类标准》（CXS 212-1999）。

全脂乳粉

乳脂肪含量	≥26%，<42%
水分最高含量 (a)	5% m/m
乳的非脂固体中乳蛋白最低含量 (a)	34% m/m

部分脱脂乳粉

乳脂肪含量	>1.5，<26
水分最高含量 (a)	5% m/m
乳的非脂固体中乳蛋白最低含量 (a)	34% m/m

脱脂乳粉

乳脂肪最高含量	1.5% m/m
水分最高含量 (a)	5% m/m
乳的非脂固体中乳蛋白最低含量 (a)	34% m/m

（a）不包括乳糖的结晶水，乳的非脂固体含量包括乳糖结晶水。

4　食品添加剂

只允许在规定范围内使用下表列出的食品添加剂。

INS编号	添加剂名称	最大限量
稳定剂		
331	柠檬酸钠	5 000 mg/kg，单用或混用，以无水物质计
332	柠檬酸钾	
固化剂		
508	氯化钾	根据GMP限量使用
509	氯化钙	
酸度调节剂		
339	磷酸钠	5 000 mg/kg，单用或混用，以无水物质计
340	磷酸钾	
450	二磷酸盐	
451	三磷酸盐	
452	聚磷酸盐	
500	碳酸钠	
501	碳酸钾	

INS编号	添加剂名称	最大限量
乳化剂		
322	卵磷脂	根据GMP限量使用
471	单、双甘油脂肪酸酯	2 500 mg/kg
消结块剂		
170（i）	碳酸钙	10 000 mg/kg，单用或混用
341（iii）	磷酸三钙	
343（iii）	磷酸三镁	
504（i）	碳酸镁	
530	氧化镁	
551	二氧化硅（无定型）	
552	硅酸钙	
553	硅酸镁	
554	硅酸铝钠	265 mg/kg，以铝计
抗氧化剂		
300	L-抗坏血酸	500 mg/kg，以抗坏血酸计
301	抗坏血酸钠	
304	抗坏血酸棕榈酸酯	
320	丁基羟基茴香醚	100 mg/kg

5 **污染物**

本标准所涉及的产品应符合《食品及饲料中污染物和毒素通用标准》（CXS 193-1995）规定的污染物最大限量。

本标准所涉及的产品在加工中使用的牛乳应符合《食品及饲料中污染物和毒素通用标准》（CXS 193-1995）规定的乳中污染物和毒素最大限量以及国际食品法典委员会设定的农药和兽药最大残留限量。

6 **卫生要求**

建议本标准所涉及的产品应遵循《食品卫生总则》（CXC 1-1969）、《乳及乳制品卫生操作规范》（CXC 57-2004）以及《卫生操作规范》和《生产操作规范》等其他相关法典文本。产品应符合《食品微生物标准制定与实施原则和准则》（CXG 21-1997）规定的所有微生物标准。

7 **标识**

除符合《预包装食品标识通用标准》（CXS 1-1985）和《乳品术语》（CXS 206-

1999）外，还应符合下列具体规定。

7.1 产品名称

产品名称应为：

— 奶油粉	
— 全脂乳粉	根据第3.2条产品成分
— 部分脱脂乳粉	
— 脱脂乳粉	

如果部分脱脂乳粉的乳脂肪含量在14%~16%（质量分数），可以命名为"半脱脂乳粉"。

如果国家法律允许或便于产品销售国的消费者识别，"全脂乳粉"可以命名为"全奶油乳粉"，"脱脂乳粉"可命名为"低脂乳粉"。

7.2 乳脂含量说明

如果省略乳脂含量可能误导消费者，则应以产品最终消费国可以接受的方式声明：（i）质量百分比形式；（ii）如果产品标识上标明了份数，也可用每份中乳脂重量（g）表示。

7.3 乳蛋白含量说明

如果省略乳蛋白含量可能误导消费者，则应以产品最终消费国可以接受的方式声明：（i）质量百分比形式；（ii）如果产品标识上标明了份数，也可用每份中乳蛋白重量（g）表示。

7.4 配料表

尽管《预包装食品标识通用标准》（CXC 1-1985）第4.2.1条有规定，但仅用于调整乳蛋白含量的乳制品不需要声明。

7.5 非零售包装标识

在包装容器上除要标明产品名称、批次、生产厂家或包装商的名称和地址外，在包装容器上或附随说明书中，也应按本标准第7条和《预包装食品标识通用标准》（CXS 1-1985）第4.1~4.8条所要求的信息以及储存说明（在必要的地方）加以陈述。如果批次、生产厂家或包装商的名称和地址可以在附随的文件标明，则可以用一个识别标识来代替。

8 抽样和分析方法

为检查产品是否符合本标准，应采用《分析和抽样推荐性方法》（CXS 234-1999）中涉及本标准规定的分析和抽样方法。

附录

附 加 信 息

以下补充信息对前文各条款的规定不构成影响,前文规定内容对产品标识、食品名称的使用以及食品安全性至关重要。

要求	全脂乳粉	部分脱脂乳粉	脱脂乳粉	方法
可滴定酸度 (0.1 mol/L NaOH mL/10 g非脂固体)	≤18.0	≤18.0	≤18.0	参见CXS 234-1999
焦化颗粒	B类以及下	B类以及下	B类以及下	参见CXS 234-1999
可溶性指数(mL)	≤1.0	≤1.0	≤1.0	参见CXS 234-1999

STANDARD FOR MILK POWDERS AND CREAM POWDER

CXS 207-1999

This Standard replaced the *Standard for Whole Milk Powder, Partly Skimmed Milk Powder and Skimmed Milk Powder* (A-5-1971) and the *Standard for Cream Powder, Half Cream Powder and High Fat Milk Powder* (A-10-1971).
Adopted in 1999. Amended in 2010, 2013, 2014, 2016, 2018.

1 SCOPE

This Standard applies to milk powders and cream powder, intended for direct consumption or further processing, in conformity with the description in Section 2 of this Standard.

2 DESCRIPTION

Milk powders and cream powder are milk products which can be obtained by the partial removal of water from milk or cream. The fat and/or protein content of the milk or cream may have been adjusted, only to comply with the compositional requirements in Section 3 of this Standard, by the addition and/or withdrawal of milk constituents in such a way as not to alter the whey protein to casein ratio of the milk being adjusted.

3 ESSENTIAL COMPOSITION AND QUALITY FACTORS

3.1 Raw materials

Milk and cream

The following milk products are allowed for protein adjustment purposes:

- Milk retentate: Milk retentate is the product obtained by concentrating milk protein by ultrafiltration of milk, partly skimmed milk, or skimmed milk;
- Milk permeate: Milk permeate is the product obtained by removing milk proteins and milkfat from milk, partly skimmed milk, or skimmed milk by ultrafiltration; and
- Lactose[1].

3.2 Composition

Cream powder

Minimum milkfat	42% m/m
Maximum water[a]	5% m/m
Minimum milk protein in milk solids-not-fat[a]	34% m/m

Whole milk powder

Milkfat	Minium 26% and less than 42% m/m
Maximum water[a]	5% m/m

---

1 See *Standard for Sugars* (CXS 212-1999).

Minimum milk protein in milk solids-not-fat[a]   34% m/m

Partly skimmed milk powder
Milkfat   More than 1.5% and less than 26% m/m
Maximum water[a]   5% m/m
Minimum milk protein in milk solids-not-fat[a]   34% m/m

Skimmed milk powder
Maximum milkfat   1.5% m/m
Maximum water[a]   5% m/m
Minimum milk protein in milk solids-not-fat[a]   34% m/m

(a) The water content does not include water of crystallization of the lactose; the milk solids-not-fat content includes water of crystallization of the lactose.

4 FOOD ADDITIVES

Only those food additives listed below may be used and only within the limits specified.

INS no.	Name of additive	Maximum level
Stabilizers		
331	Sodium citrates	5 000 mg/kg singly or in combination, expressed as anhydrous substances
332	Potassium citrates	
Firming agents		
508	Potassium chloride	Limited by GMP
509	Calcium chloride	
Acidity regulators		
339	Sodium phosphates	5 000 mg/kg singly or in combination, expressed as anhydrous substances
340	Potassium phosphates	
450	Diphosphates	
451	Triphosphates	
452	Polyphosphates	
500	Sodium carbonates	
501	Potassium carbonates	
Emulsifiers		
322	Lecithins	Limited by GMP
471	Mono- and diglycerides of fatty acids	2 500 mg/kg

INS no.	Name of additive	Maximum level
Anticaking agents		
170(i)	Calcium carbonate	10 000 mg/kg singly or in combination
341(iii)	Tricalcium phosphate	
343(iii)	Trimagnesium phosphate	
504(i)	Magnesium carbonate	
530	Magnesium oxide	
551	Silicon dioxide, amorphous	
552	Calcium silicate	
553	Magnesium silicates	
554	Sodium aluminium silicate	265 mg/kg expressed as aluminium
Antioxidants		
300	Ascorbic acid, *L*-	500 mg/kg expressed as ascorbic acid
301	Sodium ascorbate	
304	Ascorbyl palmitate	
320	Butylated hydroxyanisole	100 mg/kg

5 CONTAMINANTS

The products covered by this Standard shall comply with the Maximum Levels for contaminants that are specified for the product in the *General Standard for Contaminants and Toxins in Food and Feed* (CXS 193-1995).

The milk used in the manufacture of the products covered by this Standard shall comply with the Maximum Levels for contaminants and toxins specified for milk by the *General Standard for Contaminants* and *Toxins in Food and Feed* (CXS 193-1995) and with the maximum residue limits for veterinary drug residues and pesticides established for milk by the CAC.

6 HYGIENE

It is recommended that the products covered by the provisions of this standard be prepared and handled in accordance with the appropriate sections of the *General Principles of Food Hygiene* (CXC 1-1969), the *Code of* Hygienic Practice for Milk and Milk Products (CXC 57-2004) and other relevant Codex texts such as Codes of Hygienic Practice and Codes of Practice.

The products should comply with any microbiological criteria established in accordance with the *Principles and Guidelines for the Establishment and Application of Microbiological Criteria Related to Foods* (CXG 21-1997).

7 LABELLING

In addition to the provisions of the *General Standard for the Labelling of Prepackaged Foods* (CXS 1-1985) and the *General Standard for the Use of Dairy Terms* (CXS 206-

1999), the following specific provisions apply:

7.1 Name of the food

The name of the food shall be:

- Cream powder
- Whole milk powder
- Partly skimmed milk powder
- Skimmed milk powder

according to the composition in Section 3.2

Partly skimmed milk powder may be designated "Semi-skimmed milk powder" provided that the content of milkfat does not exceed 16% m/m and is not less than 14% m/m.

If allowed by national legislation or otherwise identified to the consumer in the country where the product is sold, "whole milk powder" may be designated "full cream milk powder" and "skimmed milk powder" may be designated "low fat milk powder".

7.2 Declaration of milkfat content

If the consumer would be misled by the omission, the milkfat content shall be declared in a manner found acceptable in the country of sale to the final consumer, either (i) as a percentage by mass, or (ii) in grams per serving as quantified in the label provided that the number of servings is stated.

7.3 Declaration of milk protein

If the consumer would be misled by the omission, the milk protein content shall be declared in a manner acceptable in the country of sale to the final consumer, either as (i) a percentage by mass, or (ii) grams per serving as quantified in the label provided the number of servings is stated.

7.4 List of ingredients

Notwithstanding the provision of Section 4.2.1 of the *General Standard for the Labelling of Prepackaged Foods* (CXS 1-1985), milk products used only for protein adjustment need not be declared.

7.5 Labelling of non-retail containers

Information required in Section 7 of this Standard and Sections 4.1 to 4.8 of the *General Standard for the* Labelling *of Prepackaged Foods* (CXS 1-1985), and, if necessary, storage instructions, shall be given either on the container or in accompanying documents, except that the name of the product, lot identification, and the name and address of the manufacturer or packer shall appear on the container. However, lot identification, and the name and address of the manufacturer or packer may be replaced by an identification mark, provided that such a mark is clearly identifiable with the accompanying documents.

8 METHODS OF SAMPLING AND ANALYSIS

For checking the compliance with this standard, the methods of analysis and sampling contained in the *Recommended Methods of Analysis and Sampling* (CXS 234-1999) relevant to the provisions in this standard, shall be used.

APPENDIX

ADDITIONAL INFORMATION

The additional information below does not affect the provisions in the preceding sections which are those that are essential to the product identity, the use of the name of the food and the safety of the food.

Additional quality factors	Whole milk powder	Partially skimmed milk powder	Skimmed milk powder	Method
Titratable acidity	max 18.0	max 18.0	max 18.0	See CXS 234-1999
(mL-0.1 N NaOH/ 10 g-solids-not-fat)				See CXS 234-1999
Scorched particles	max Disc B	max Disc B	max Disc B	See CXS 234-1999
Solubility index (mL)	max 1.0	max 1.0	max 1.0	See CXS 234-1999

盐水干酪

GROUP STANDARD FOR CHEESES IN BRINE

CXS 208-1999

1999年通过。2001年、2010年、2018年修订。

1 范围

本标准适用于符合本标准第2条所述即食或用于进一步加工的盐水干酪,从属于本类标准的国际食品法典标准个别盐水干酪品种可能更具体,但那些特定的规定应受本标准约束。

2 说明

盐水干酪是指符合《干酪通则》(CXS 283-1978)的半硬质到软质的成熟干酪。该干酪本身呈白色至浅黄色,质地坚硬,适合切片,几乎不能进行机械操作。干酪无真正的外皮,在供给消费者或进行预包装之前一直在盐水中成熟和保存。个别盐水干酪品种含有特定的香草和香料作为其特性组成部分。

3 基本成分和质量要求

3.1 基本原料

乳和/或乳制品。

3.2 其他配料

- 无害乳酸发酵剂和/或产香菌和其他无害微生物的培养物;
- 安全、适宜的酶;
- 氯化钠;
- 饮用水;
- 能体现盐水干酪部分特性的香草和香料。

3.3 成分

	软质	半硬质
干物质中最低脂肪含量(%)	40	40
最低干物质含量(%)	40	52

4 食品添加剂

只允许在规定范围内使用下表列出的食品添加剂。

INS编号	添加剂名称	最大限量
酸度调节剂		
270	乳酸（L-，D-和DL-）	根据GMP限量使用
575	葡萄糖酸-δ-内酯	根据GMP限量使用

5 污染物

本标准所涉及的产品应符合《食品及饲料中污染物和毒素通用标准》（CXS 193-1995）规定的污染物最大限量。

本标准所涉及的产品在加工中使用的牛乳应符合《食品及饲料中污染物和毒素的通用标准》（CXS 193-1995）规定的乳中污染物和毒素最大限量以及国际食品法典委员会设定的农药和兽药最大残留限量。

6 卫生要求

建议本标准所涉及的产品应遵循《食品卫生总则》（CXC 1-1969）、《乳及乳制品卫生操作规范》（CXC 57-2004）以及《卫生操作规范》和《生产操作规范》等其他相关法典文本。产品应符合《食品微生物标准制定与实施原则和准则》（CXG 21-1997）规定的所有微生物学标准。

7 标识

除符合《预包装食品标识通用标准》（CXS 1-1985）和《乳品术语》（CXS 206-1999）外，还应符合下列具体规定。

7.1 产品名称

产品名称应为盐水干酪。然而，在食典标准中确定的某一特定盐水干酪的专有名称中，"盐水干酪"一词可以省略，在省略的情况下，其品名应是产品销售国法律指定名称，而且该省略不会使消费者对产品特性产生误解。

7.2 乳脂含量说明

乳脂含量应以销售国可接受的方式向最终消费者声明：（i）质量百分比形式；（ii）干物质中脂肪百分比形式；（iii）如果产品标识上标明了份数，也可用每份中乳脂重量（g）表示。

另外，可以使用下列术语：

高脂	（如果干物质脂肪含量≥60%）
全脂	（如果干物质脂肪含量≥45%，<60%）
中脂	（如果干物质脂肪含量≥25%，<45%）
部分脱脂	（如果干物质脂肪含量≥10%，<25%）
脱脂	（如果干物质脂肪含量<10%）

7.3 非零售包装标识

在包装容器上除要标明产品名称、批次、生产厂家或包装商的名称和地址外，在包装容器上或附随说明书中，也应按本标准第7条和《预包装食品标识通用标准》（CXS 1-1985）第4.1～4.8条所要求的信息以及储存说明（在必要的地方）加以陈述。如果批次、生产厂家或包装商的名称和地址可以在附随的文件标明，则可以用一个识别标识来代替。

8　抽样和分析方法

为检查产品是否符合本标准，应采用《分析和抽样推荐性方法》（CXS 234-1999）中涉及本标准规定的分析和抽样方法。

8.1　抽样

盐水干酪抽样方法：把干酪样品放到布上或吸水纸上5～10 min，切下2～3 cm薄片放入密封盒内送交实验室分析。

GROUP STANDARD FOR CHEESES IN BRINE

CXS 208-1999

Adopted in 1999. Amended in 2001, 2010, 2018.

1 SCOPE

This Standard applies to Cheeses in Brine, intended for direct consumption or further processing, in conformity with the description in Section 2 of this Standard. Subject to the provisions of this Group Standard, Codex standards for individual varieties of Cheeses in Brine may contain provisions which are more specific than those in this Standard.

2 DESCRIPTION

Cheeses in Brine are semi-hard to soft ripened cheeses in conformity with the *General Standard for Cheese* (CXS 283-1978). The body has a white to yellowish colour and a compact texture suitable for slicing, with none to few mechanical openings. The cheeses have no actual rind and have been ripened and preserved in brine until delivered to, or prepacked for, the consumer. Certain individual Cheeses in Brine contain specific herbs and spices as part of their identity.

3 ESSENTIAL COMPOSITION AND QUALITY FACTORS

3.1 Raw materials

Milk and/or products obtained from milk.

3.2 Permitted ingredients

- Starter cultures of harmless lactic acid and/or flavour producing bacteria and cultures of other harmless micro-organisms;
- Safe and suitable enzymes;
- Sodium chloride;
- Potable water;
- Herbs and spices where part of the identity of the Cheese in Brine.

3.3 Composition

	Soft	Semi-hard
Minimum fat in dry matter(%)	40	40
Minimum dry matter(%)	40	52

4 FOOD ADDITIVES

Only those food additives listed may be used and only within the limits specified.

INS no.	Name of additive	Maximum level
Acidity regulators		
270	Lactic acid, *L*-, *D*- and *DL*-	Limited by GMP
575	Glucono *delta*-lactone	Limited by GMP

5 CONTAMINANTS

The products covered by this Standard shall comply with the maximum levels for contaminants that are specified for the product in the *General Standard for Contaminants and Toxins in Food and Feed* (CXS 193-1995).

The milk used in the manufacture of the products covered by this Standard shall comply with the maximum levels for contaminants and toxins specified for milk by the *General Standard for Contaminants and Toxins in Food and Feed* (CXS 193-1995) and with the maximum residue limits for veterinary drug residues and pesticides established for milk by the CAC.

6 HYGIENE

It is recommended that the products covered by the provisions of this Standard be prepared and handled in accordance with the appropriate sections of the *General Principles of Food Hygiene* (CXC 1-1969), the *Code of Hygienic Practice for Milk and Milk Products* (CXC 57-2004) and other relevant Codex texts such as Codes of Hygienic Practice and Codes of Practice. The products should comply with any microbiological criteria established in accordance with the *Principles and Guidelines for the Establishment and Application of Microbiological Criteria Related to Foods* (CXG 21-1997).

7 LABELLING

In addition to the provisions of the *General Standard for the Labelling of Prepackaged Foods* (CXS 1-1985) and the *General Standard for the Use of Dairy Terms* (CXS 206-1999), the following specific provisions apply.

7.1 Name of the food

The name of the food shall be Cheese in Brine. However, the word "Cheese in Brine" may be omitted in the designation of an individual Cheese in Brine variety reserved by a Codex standard for individual Cheese in Brine, and, in the absence thereof, a variety name specified in the national legislation of the country in which the product is sold, provided that the omission does not create an erroneous impression regarding the character of the food.

7.2 Declaration of milkfat content

The milkfat content shall be declared in a manner found acceptable in the country of sale to the final consumer, either (i) as a percentage by mass, (ii) as a percentage of fat in dry matter, or (iii) in grams per serving as quantified in the label provided that the number of servings is stated.

Additionally, the following terms may be used:

High fat	(if the content of FDM is above or equal to 60%)
Full fat	(if the content of FDM is above or equal to 45% and less than 60%)
Medium fat	(if the content of FDM is above or equal to 25% and less than 45%)
Partially skimmed	(if the content of FDM is above or equal to 10% and less than 25%)
Skim	(if the content of FDM is less than 10%)

7.3 Labelling of non-retail containers

Information required in Section 7 of this Standard and Sections 4.1 to 4.8 of the *General Standard for the Labelling of Prepackaged Foods* (CXS 1-1985), and, if necessary, storage instructions, shall be given either on the container or in accompanying documents, except that the name of the product, lot identification, and the name and address of the manufacturer or packer shall appear on the container. However, lot identification, and the name and address of the manufacturer or packer may be replaced by an identification mark, provided that such a mark is clearly identifiable with the accompanying documents.

8 METHODS OF SAMPLING AND ANALYSIS

For checking the compliance with this Standard, the methods of analysis and sampling contained in the *Recommended Methods of Analysis and Sampling* (CXS 234-1999) relevant to the provisions in this Standard, shall be used.

8.1 Sampling

Special requirements for Cheese in Brine: A representative piece of cheese is placed on a cloth or on a sheet of absorbent paper for 5 to 10 min. A slice of 2–3 cm is cut off and sent to the laboratory in a sealed insulated box for analysis.

未成熟干酪（包括新鲜干酪）

GROUP STANDARD FOR UNRIPENED CHEESE
INCLUDING FRESH CHEESE

CXS 221-2001

2001年通过。2008年、2010年、2013年、2018年修订

1　范围

本标准适用于符合本标准第2条所述即食或用于进一步加工的未成熟干酪（包括新鲜干酪），从属于本标准规定的未成熟干酪品种在国际食品法典标准的规定可能比本标准更具体，应该遵循国际食品法典标准中的具体规定。

2　说明

未成熟干酪（包括新鲜干酪）是指在制作后短期内即食的符合《干酪通则》（CXS 283-1978）的产品。

3　基本成分和质量要求

3.1　基本原料

乳和/或乳制品。

3.2　其他配料

- 无害乳酸发酵剂和/或产香菌和其他无害微生物的培养剂；
- 凝乳酶或其他安全、适宜的凝固酶；
- 氯化钠；
- 饮用水；
- 明胶和淀粉：在《干酪通则》（CXS 283-1978）规定中，这些物质能够与稳定剂起相同作用，但仅能添加《良好操作规范》所规定的作用用量同时考虑第4条列出的稳定剂/增稠剂用途；
- 醋；
- 大米、玉米、马铃薯粉和淀粉：尽管有《干酪通则》（CXS 283-1978），但这些物质同样可作为抗结块剂处理切块、切片和切碎产品表面，只要添加量是良好操作规范所要求达到期待功能的必需量，同时考虑第4条规定的抗结块剂的使用。

4 食品添加剂

只允许在规定范围内使用下表列出的食品添加剂。下面未列出但在个别不同未成熟干酪的国际食品法典标准中提出的添加剂，在规定限度内，也可用于类似干酪类型。

INS编号	添加剂名称	最大限量
酸度调节剂		
170	碳酸钙	根据GMP限量使用
260	冰醋酸	
270	乳酸（L-，D-和DL-）	
296	DL-苹果酸	
330	柠檬酸	
338	磷酸	880 mg/kg，以磷计
500	碳酸钠	根据GMP限量使用
501	碳酸钾	
507	盐酸	
575	葡萄糖酸-δ-内酯	
稳定剂/增稠剂		
符合乳制品定义的产品可以使用稳定剂和增稠剂（包括改性淀粉），且仅在它们是功能上必须的时候可以使用并考虑到第3.2条提到的明胶和淀粉的使用。		
331	柠檬酸钠	根据GMP限量使用
332	柠檬酸钾	
333	柠檬酸钙	
339	磷酸钠	1 540 mg/kg，单独或混合，以磷计
340	磷酸钾	
341	磷酸钙	
450（i）	磷酸二钠	
450（ii）	焦磷酸三钠	
400	海藻酸	根据GMP限量使用
401	海藻酸钠	
402	海藻酸钾	
403	海藻酸铵	
404	海藻酸钙	根据GMP限量使用

INS编号	添加剂名称	最大限量
405	海藻酸丙二醇酯	5 g/kg
406	琼脂	根据GMP限量使用
407	卡拉胶	
410	槐豆胶（又名刺槐豆胶）	
412	瓜尔胶	
413	黄蓍胶	
415	黄原胶	
416	刺梧桐胶	
417	刺云实胶	
440	果胶	
460	纤维素	
466	羧甲基纤维素钠（纤维素胶）	
576	葡萄糖酸钠	
改性淀粉：		
1400	糊精（烤淀粉白色和黄色）	根据GMP限量使用
1401	酸处理淀粉	
1402	碱处理淀粉	
1403	漂白淀粉	
1404	氧化淀粉	
1405	酶处理淀粉	
1410	单淀粉磷酸酯	
1412	三偏磷酸钠醋化的磷酸双淀粉；三氯氧磷醋化的磷酸双淀粉	
1413	磷酸化二淀粉磷酸酯	
1414	乙酰化二淀粉磷酸酯	
1420	醋酸酯淀粉	
1422	乙酰化双淀粉己二酸酯	
1440	羟丙基淀粉	
1442	羟丙基二淀粉磷酸酯	

INS编号	添加剂名称	最大限量
着色剂		
100	姜黄素（用于可食的干酪皮）	根据GMP限量使用
101	核黄素	
140	叶绿素	
141	叶绿素铜	15 mg/kg，单用或混用
160a（i）	β-胡萝卜素（人工合成）	25 mg/kg
160a（ii）	β-胡萝卜素（天然萃取物）	600 mg/kg
160b（ii）	胭脂树提取物-红木素	25 mg/kg
160c	辣椒油树脂	根据GMP限量使用
160e	β-apo-8'-胡萝卜素	35 mg/kg
160f	β-apo-8'-胡萝卜素酸乙酯	
162	甜菜红	根据GMP限量使用
171	二氧化钛	
防腐剂		
200	山梨酸	1 000 mg/kg，单独或混用，以山梨酸计
202	山梨酸钾	
203	山梨酸钙	
234	乳酸链球菌素	12.5 mg/kg
280	丙酸	根据GMP限量使用
281	丙酸钠	
282	丙酸钙	
283	丙酸钾	
仅用于表面/外皮处理：		
235	纳他霉素（匹马菌素）	表面2 mg/dm^2，在5 mm深度处不存在
发泡剂（仅用于攒奶油的产品）		
290	二氧化碳	根据GMP限量使用
941	氮气	
抗结块剂［仅切片、切块、切碎和磨碎的产品（表面处理）］		
460	纤维素	根据GMP限量使用

INS编号	添加剂名称	最大限量
551	二氧化硅（无定型）	10 000 mg/kg，单独或混用，以二氧化硅计算硅盐
552	硅酸钙	
553	硅酸镁	
560	硅酸钾	
防腐剂［仅切片、切块、切碎和磨碎的产品（表面处理）］		
200	山梨酸	1 000 mg/kg，单用或混用，以山梨酸计
202	山梨酸钾	
203	山梨酸钙	
280	丙酸	根据GMP限量使用
281	丙酸钠	
282	丙酸钙	
283	丙酸钾	
235	纳他霉素（匹马菌素）	20 mg/kg（在揉搓和伸拉过程中用于表面）

5 污染物

本标准所涉及的产品应符合《食品及饲料中污染物和毒素的通用标准》（CXS 193-1995）规定的污染物最大限量。

本标准所涉及的产品在加工中使用的牛乳应符合《食品及饲料中污染物和毒素的通用标准》（CXS 193-1995）规定的乳中污染物和毒素最大限量以及国际食品法典委员会设定的农药和兽药最大残留限量。

6 卫生要求

建议本标准所涉及的产品应遵循《食品卫生总则》（CXC 1-1969）、《乳及乳制品卫生操作规范》（CXC 57-2004）以及《卫生操作规范》和《生产操作规范》等其他相关法典文本。产品应符合《食品微生物标准制定与实施原则和准则》（CXG 21-1997）规定的所有微生物标准。

7 标识

除符合《预包装食品标识通用标准》（CXS 1-1985）和《乳品术语》（CXS 206-1999）外，还应符合下列具体规定。

7.1 产品名称

产品名称应为未成熟干酪。如果某个未成熟干酪品种已被收录在未成熟干酪的国际食品法典标准中，且在产品销售国法律中已对此产品指定了具体名称，只要省略不会使消费者对产品特性产生误解，命名可忽略"未成熟干酪"一词。

在产品未被另外命名但指定为"未成熟干酪",这种名称应伴随一个在《干酪通则》(CXS 283-1978)的第7.1.1条中提供的描述性术语。

只要不会误导产品销售国消费者,未成熟干酪也可以命名为"新鲜干酪"。

7.2 乳脂含量说明

乳脂含量应以销售国可接受的方式向最终消费者声明:(i)质量百分比形式;(ii)干物质中脂肪百分比形式;(iii)如果产品标识上标明了份数,也可用每份中乳脂重量(g)表示。

另外,可以使用下列术语:

高脂	(如果干物质脂肪含量≥60%)
全脂	(如果干物质脂肪含量≥45%,<60%)
中脂	(如果干物质脂肪含量≥25%,<45%)
部分脱脂	(如果干物质脂肪含量≥10%,<25%)
脱脂	(如果干物质脂肪含量<10%)

7.3 非零售包装标识

在包装容器上除要标明产品名称、批次、生产厂家或包装商的名称和地址外,在包装容器上或附随说明书中,也应按本标准第7条和《预包装食品标识通用标准》(CXS 1-1985)第4.1~4.8条所要求的信息以及储存说明(在必要的地方)加以陈述。如果批次、生产厂家或包装商的名称和地址可以在附随的文件标明,则可以用一个识别标识代替。

8 抽样和分析方法

为检查产品是否符合本标准,应采用《分析和抽样推荐性方法》(CXS 234-1999)中涉及本标准规定的分析和抽样方法。

GROUP STANDARD FOR UNRIPENED CHEESE INCLUDING FRESH CHEESE

CXS 221-2001

Adopted in 2001. Amended in 2008, 2010, 2013, 2018.

1 SCOPE

This Standard applies to unripened cheese including fresh cheese, intended for direct consumption or further processing, in conformity with the description in Section 2 of this Standard. Subject to the provisions of this Standard, Codex Standards for individual varieties of unripened cheese may contain provisions, which are more specific than those in this Standard and in these cases; those specific provisions shall apply.

2 DESCRIPTION

Unripened cheeses including fresh cheeses are products in conformity with the *General Standard for Cheese* (CXS 283-1978), which are ready for consumption shortly after manufacture.

3 ESSENTIAL COMPOSITION AND QUALITY FACTORS

3.1 Raw materials

Milk and/or products obtained from milk.

3.2 Permitted ingredients

- Starter cultures of harmless lactic acid and/or flavour producing bacteria and cultures of other harmless micro-organisms;
- Rennet or other safe and suitable coagulating enzymes;
- Sodium chloride;
- Potable water;
- Gelatine and starches: Notwithstanding the provisions in the *General Standard for Cheese* (CXS 283-1978), these substances can be used in the same function as stabilizers, provided they are added only in amounts functionally necessary as governed by Good Manufacturing Practice taking into account any use of the stabilisers/thickeners listed in Section 4;
- Vinegar;
- Rice, corn and potato flours and starches: Notwithstanding the provisions in the *General Standard for Cheese* (CXS 283-1978), these substances can be used in the same function as anti-caking agents for treatment of the surface of cut, sliced, and shredded products only, provided they are added only in amounts functionally necessary as governed by Good Manufacturing Practice taking into account any use of the anti-caking agents listed in Section 4.

4 FOOD ADDITIVES

Only those food additives listed below may be used and only within the limits specified.

Additives not listed below but provided for in individual Codex standards for varieties of Unripened Cheeses may also be used in similar types of cheese within the limits specified within those standards.

INS no.	Name of additive	Maximum level
Acidity regulators		
170	Calcium carbonates	Limited by GMP
260	Acetic acid, glacial	
270	Lactic acid, *L*-, *D*- and *DL*-	
296	Malic acid, *DL*-	
330	Citric acid	
338	Phosphoric acid	880 mg/kg expressed as phosphorous
500	Sodium carbonates	Limited by GMP
501	Potassium carbonates	
507	Hydrochloric acid	
575	Glucono *delta*-lactone (GDL)	
Stabilizers/thickeners		
Stabilizers and thickeners including modified starches may be used in compliance with the definition for milk products and only to the extent they are functionally necessary taking into account any use of gelatine and starch as provided for in Section 3.2.		
331	Sodium citrates	Limited by GMP
332	Potassium citrates	
333	Calcium citrates	
339	Sodium phosphates	1 540 mg/kg, singly or in combination, expressed as phosphorous
340	Potassium phosphates	
341	Calcium phosphates	
450(i)	Disodium diphosphate	
450(ii)	Trisodium diphosphate	
400	Alginic acid	Limited by GMP
401	Sodium alginate	
402	Potassium alginate	
403	Ammonium alginate	

GROUP STANDARD FOR UNRIPENED CHEESE INCLUDING FRESH CHEESE

INS no.	Name of additive	Maximum level
404	Calcium alginate	Limited by GMP
405	Propylene glycol alginate	5 g/kg
406	Agar	Limited by GMP
407	Carrageenan	
410	Carob bean gum	
412	Guar gum	
413	Tragacanth gum	
415	Xanthan gum	
416	Karaya gum	
417	Tara gum	
440	Pectins	
460	Cellulose	
466	Sodium carboxymethyl cellulose (Cellulose gum)	
576	Sodium gluconate	
Modified starches as follows:		
1400	Dextrins, roasted starch white and yellow	Limited by GMP
1401	Acid-treated starch	
1402	Alkaline treated starch	
1403	Bleached starched	
1404	Oxidized starch	
1405	Starches, enzyme-treated	
1410	Monostarch phosphate	
1412	Distarch phosphate esterified with sodium trimetasphosphate; esterified with phosphorus oxychloride	
1413	Phosphated distarch phosphate	
1414	Acetylated distarch phosphate	
1420	Starch acetate	
1422	Acetylated distarch adipate	
1440	Hydroxypropyl starch	
1442	Hydroxypropyl distarch phosphate	

INS no.	Name of additive	Maximum level
Colours		
100	Curcumins (*for edible cheese rind*)	Limited by GMP
101	Riboflavins	
140	Chlorophyll	
141	Copper chlorophylls	15 mg/kg, singly or combined
160a(i)	Carotene, *beta*-, synthetic	25 mg/kg
160a(ii)	Carotenes, *beta*-, vegetable	600 mg/kg
160b(ii)	Annatto extracts – norbixin based	25 mg/kg
160c	Paprika oleoresins	Limited by GMP
160e	Carotenal, *beta*-apo-8'-	35 mg/kg
160f	Carotenoic acid, ethyl ester, *beta*-apo-8'-	
162	Beet red	Limited by GMP
171	Titanium dioxide	
Preservatives		
200	Sorbic acid	1 000 mg/kg of cheese, singly or in combination, expressed as sorbic acid
202	Potassium sorbate	
203	Calcium sorbate	
234	Nisin	12.5 mg/kg
280	Propionic acid	Limited by GMP
281	Sodium propionate	
282	Calcium propionate	
283	Potassium propionate	
For surface/rind treatment only:		
235	Natamycin (pimaricin)	2 mg/dm^2 of surface. Not present in a depth of 5 mm.
Foaming agents (for whipped products only)		
290	Carbon dioxide	Limited by GMP
941	Nitrogen	
Anticaking agents (Sliced, cut, shredded and grated products only (surface treatment))		
460	Cellulose	Limited by GMP

INS no.	Name of additive	Maximum level
551	Silicon dioxide, amorphous	10 000 mg/kg singly or in combination. Silicates calculated as silicon dioxide
552	Calcium silicate	
553	Magnesium silicates	
560	Potassium silicate	
Preservatives (Sliced, cut, shredded and grated products only (surface treatment))		
200	Sorbic acid	1 000 mg/kg of cheese, singly or in combination, expressed as sorbic acid
202	Potassium sorbate	
203	Calcium sorbate	
280	Propionic acid	Limited by GMP
281	Sodium propionate	
282	Calcium propionate	
283	Potassium propionate	
235	Natamycin (pimaricin)	20 mg/kg applied to the surface added during kneading and stretching process

5 CONTAMINANTS

The products covered by this Standard shall comply with the maximum levels for contaminants that are specified for the product in the *General Standard for Contaminants and Toxins in Food and Feed* (CXS 193-1995).

The milk used in the manufacture of the products covered by this Standard shall comply with the maximum levels for contaminants and toxins specified for milk by the *General Standard for Contaminants and Toxins in Food and Feed* (CXS 193-1995) and with the maximum residue limits for veterinary drug residues and pesticides established for milk by the CAC.

6 HYGIENE

It is recommended that the products covered by the provisions of this Standard be prepared and handled in accordance with the appropriate sections of the *General Principles of Food Hygiene* (CXC 1-1969), the *Code of Hygienic Practice for Milk and Milk Products* (CXC 57-2004) and other relevant Codex texts such as Codes of Hygienic Practice and Codes of Practice. The products should comply with any microbiological criteria established in accordance with the *Principles and Guidelines for the Establishment and Application of Microbiological Criteria Related to Foods* (CXG 21-1997).

7 LABELLING

In addition to the provisions of the *General Standard for the Labelling of Prepackaged Foods* (CXS 1-1985) and the *General Standard for the Use of Dairy Terms* (CXS 206-1999), the following specific provisions apply.

7.1 **Name of the food**

The name of the food shall be unripened cheese. However, the words "unripened cheese" may be omitted in the designation of an individual unripened cheese variety reserved by a Codex standard for individual cheeses, and, in the absence thereof, a variety name specified in the national legislation of the country in which the product is sold, provided that the omission does not create an erroneous impression regarding the character of the food.

In case the product is not designated by an alternative or a variety name, but with the designation "unripened cheese", the designation may be accompanied by a descriptive term such as provided for in Section 7.1.1 of the *General Standard for Cheese* (CXS 283-1978).

Unripened cheese may alternatively be designated "fresh cheese" provided it is not misleading to the consumer in the country in which the product is sold.

7.2 **Declaration of milkfat content**

The milk fat content shall be declared in a manner found acceptable in the country of sale to the final consumer, either (i) as a percentage by mass, (ii) as a percentage of fat in dry matter, or (iii) in grams per serving as quantified in the label, provided that the number of servings is stated.

Additionally, the following terms may be used:

High fat	(if the content of FDM is above or equal to 60%)
Full fat	(if the content of FDM is above or equal to 45% and less than 60%)
Medium fat	(if the content of FDM is above or equal to 25% and less than 45%)
Partially skimmed	(if the content of FDM is above or equal to 10% and less than 25%)
Skim	(if the content of FDM is less than 10%)

7.3 **Labelling of non-retail containers**

Information required in Section 7 of this Standard and Sections 4.1 to 4.8 of the *General Standard for the Labelling of Prepackaged Foods* (CXS 1-1985) and, if necessary, storage instructions, shall be given either on the container or in accompanying documents, except that the name of the product, lot identification and the name and address of the manufacturer or packer shall appear on the container, and, in the absence of such a container on the cheese itself. However, lot identification and the name and address of the manufacturer or packer may be replaced by an identification mark, provided that such mark is clearly identifiable with the accompanying documents.

8 **METHODS OF SAMPLING AND ANALYSIS**

For checking the compliance with this standard, the methods of analysis and sampling contained in the *Recommended Methods of Analysis and Sampling* (CXS 234-1999) relevant to the provisions in this standard, shall be used.

发酵乳制品

STANDARD FOR FERMENTED MILKS

CXS 243-2003

2003年通过。2008年、2010年、2018年修订。

1 范围

本标准适用于符合本标准第2条所述即食或用于进一步加工的发酵乳,包括热处理发酵乳、浓缩发酵乳以及基于此类产品的调制乳制品。

2 说明

2.1 发酵乳是指乳发酵产品,由经或未经改变的乳品成分制成(符合第3.3条规定),通过特征菌作用使pH值下降,伴有或不伴凝集作用(等电沉淀)。在保质期内,产品中这些发酵微生物应该具有活性且数量丰富。如果该产品在发酵后经过热处理,则不能要求微生物必须具有活性。

发酵乳的特点取决于下列特定用于发酵的发酵剂:

酸乳:	嗜热链球菌和德氏乳杆菌保加利亚亚种混合而成的共生发酵剂
混合发酵酸乳:	嗜热链球菌和任何乳杆菌混合而成的发酵剂
嗜酸乳:	嗜酸乳杆菌
开菲尔:	由开菲尔粒、开菲尔乳杆菌、明串珠菌属、乳球菌属和乙酸杆菌属按照特定比例制备的发酵剂 开菲尔粒由乳糖酵母(马克思克鲁斯酵母)和非乳糖酵母(单孢酵母、啤酒酵母和子囊酵母)组成
马奶酒:	保加利亚乳杆菌和马克思克鲁斯酵母

也可能添加除以上特定的发酵剂以外的其他菌种。

2.2 浓缩发酵乳是指在发酵前后,将发酵乳的蛋白质浓度增加到不低于5.6%的发酵乳品。浓缩发酵乳包括传统产品,例如Stragis(去乳清酸乳)、Labneh、Ymer和Ylette。

2.3 风味发酵乳是指调制乳制品,根据《乳品术语》(CXS 206-1999)第2.3条,最多含有50%(m/m)非乳成分(如营养、非营养甜味剂,水果和蔬菜以及果汁、果浆、果肉及其制成的制剂和果酱,谷类、蜂蜜、巧克力、果仁、咖啡、香料及其他无害的天然调味食品）和/或调味剂。非乳成分可以在发酵前后添加。

2.4 发酵乳饮料是指调制乳饮品，根据《乳品术语》（CXS 206-1999）第2.3条，由2.1条所述混合发酵乳和饮用水获得，可添加或不添加乳清、其他非乳制品和调味剂等其他成分。发酵乳饮料至少含有40%（m/m）发酵乳。

除构成上述特殊发酵剂的菌种外，其他菌种也可以添加。

3 基本成分和质量要求

3.1 基本原料

- 乳和/或乳制品；
- 在复原或再制时所使用食物饮用水。

3.2 其他配料

- 无害菌种发酵剂，包括第2条所述的菌种；
- 其他合适的无害菌种（在第2.4条所述产品中）；
- 氯化钠；
- 在第2.3条（风味发酵乳）列出的非乳成分；
- 饮用水（2.4部分涉及的产品）；
- 乳及乳制品（2.4部分涉及的产品）；
- 白明胶和淀粉，在以下产品中：
  - 发酵后热处理的发酵乳；
  - 风味发酵乳；
  - 发酵乳饮料；
  - 纯发酵乳，若最终产品销售国国家法律允许。

以上配料添加仅为功能需要，根据GMP管理、生产需要适量使用。考虑第4条列出的稳定剂/增稠剂的用途，这些物质可以在加入非乳成分前后添加。

3.3 成分

	发酵乳	酸乳、混合发酵酸乳和嗜酸乳	开菲尔	马奶酒
乳蛋白质[a]（% m/m）	≥2.7%	≥2.7%	≥2.7%	
乳脂肪（% m/m）	≤10%	≤15%	≤10%	≤10%
可滴定酸度，以乳酸表示（% m/m）	≥0.3%	≥0.6%	≥0.6%	≥0.7%
乙醇（% vol./w）				≥0.5%

	发酵乳	酸乳、混合发酵酸乳和嗜酸乳	开菲尔	马奶酒
在第2.1条中定义的发酵剂微生物总数（cfu/g，总计）	$\geqslant 10^7$	$\geqslant 10^7$	$\geqslant 10^7$	$\geqslant 10^7$
标示的微生物总数[b]（cfu/g，总计）	$\geqslant 10^6$	$\geqslant 10^6$		
酵母（cfu/g）			$\geqslant 10^4$	$\geqslant 10^4$

（a）蛋白质含量为凯氏定氮法测定的总氮量乘以6.38。
（b）适用于标签中注明将某种特定微生物（第2.1条中针对所涉及产品规定的菌种除外）作为特定发酵剂补充剂添加的情况。

在风味发酵乳和发酵乳饮料中，上述标准应用于其发酵乳的部分。微生物标准（根据发酵乳所占的比例）适用于保质期到期前。这项要求不能用于发酵后热处理的产品。

产品在标签规定的储存条件下储存后，应在保质期内对产品进行分析测试，以验证是否符合上述微生物标准。

3.4 基本加工特性

在发酵乳生产过程中，除浓缩发酵乳外（第2.2条），不允许在发酵后去掉乳清。

4 食品添加剂

下表指定的添加剂类别可用于指定的产品类别。根据下表允许，每个添加剂类别仅可使用单个添加剂，并以规定范围为限。

根据《食品添加剂通用标准》（CXS 192-1995）第4.1条，风味发酵乳和发酵乳饮料中可能含有由非乳配料带入的其他添加剂。

添加剂类别	发酵乳及发酵乳饮料		热处理发酵乳及热处理发酵乳饮料	
	原味	调味	原味	调味
酸度调节剂	-	×	×	×
碳酸化剂	×[b]	×[b]	×[b]	×[b]
着色剂	-	×	-	×
乳化剂	-	×	-	×
增味剂	-	×	-	×
包装气体	-	×	×	×
防腐剂	-	-	-	×
稳定剂	×[a]	×	×	×

添加剂类别	发酵乳及发酵乳饮料		热处理发酵乳及热处理发酵乳饮料	
	原味	调味	原味	调味
甜味剂	–	×	–	×
增稠剂	×(a)	×	×	×

（a）如果产品被最终销售国家的法律准许销售，也仅限使用于重建和重组。
（b）碳酸化剂仅在发酵乳饮品中使用具备技术合理性。
× 使用该类别添加剂符合技术要求。
– 使用该类别添加剂不符合技术要求。

《食品添加剂通用标准》（CXS 192-1995）表3列出的酸度调节剂、着色剂、乳化剂、包装气体和防腐剂可用于上表列出的发酵乳产品。

INS编号	添加剂名称	最大限量
酸度调节剂		
334	$L(+)$-酒石酸	2 000 mg/kg，以酒石酸计
335（ii）	$L(+)$-酒石酸钠	
337	$L(+)$-酒石酸钾钠	
355	己二酸	1 500 mg/kg，以己二酸计
356	己二酸钠	
357	己二酸钾	
359	己二酸铵	
碳酸化剂		
290	二氧化碳	根据GMP限量使用
着色剂		
100（i）	姜黄素	100 mg/kg
101（i）	核黄素（合成物）	300 mg/kg
101（ii）	核黄素-5′-磷酸钠	300 mg/kg
102	柠檬黄	
104	喹啉黄	150 mg/kg
110	日落黄FCF	300 mg/kg
120	胭脂红	150 mg/kg
122	偶氮玉红（酸性红）	
124	丽春红4R（胭脂红A）	
129	诱惑红AC	300 mg/kg

INS编号	添加剂名称	最大限量
132	靛蓝	100 mg/kg
133	亮蓝FCF	150 mg/kg
141（i）	叶绿素，铜络合物	500 mg/kg
141（ii）	叶绿素，铜络合物，钠和钾盐	
143	固绿FCF	100 mg/kg
150b	焦糖色Ⅱ-苛性硫酸盐法	150 mg/kg
150c	焦糖Ⅲ-氨法	2 000 mg/kg
150d	焦糖Ⅳ-亚硫酸铵法	2 000 mg/kg
151	亮黑（黑色PN）	150 mg/kg
155	棕色HT	150 mg/kg
160a（i）	β-胡萝卜素（人工合成）	100 mg/kg
160e	β-apo-8′-胡萝卜素	
160f	β-apo-8′-胡萝卜酸甲酯或乙酯	
160a（iii）	β-胡萝卜素（三孢布拉霉）	
160a（ii）	β-胡萝卜素（天然萃取物）	600 mg/kg
160b（i）	胭脂树提取物-胭脂素	20 mg/kg，以胭脂素计
160b（ii）	胭脂树提取物-红木素	20 mg/kg，以红木素计
160d	番茄红素	30 mg/kg，以纯番茄红素计
161b（i）	叶黄素（万寿菊来源）	150 mg/kg
161h（i）	玉米黄质（合成物）	150 mg/kg
163（ii）	葡萄皮红	100 mg/kg
172（i）	氧化铁（黑色）	100 mg/kg
172（ii）	氧化铁（红色）	
172（iii）	氧化铁（黄色）	
乳化剂		
432	聚氧乙烯（20）山梨醇酐单月桂酸酯	3 000 mg/kg
433	聚氧乙烯（20）山梨醇酐单油酸酯	
434	聚氧乙烯（20）山梨醇酐单棕榈酸酯	
435	聚氧乙烯（20）山梨醇酐单硬脂酸酯	
436	聚氧乙烯（20）山梨醇酐三硬脂酸酯	

INS编号	添加剂名称	最大限量
472e	二乙酰酒石酸酯和脂肪酸酯	10 000 mg/kg
473	蔗糖脂肪酸酯	5 000 mg/kg
474	蔗糖甘油酯	5 000 mg/kg
475	聚甘油脂肪酸酯	2 000 mg/kg
477	脂肪酸丙二醇酯	5 000 mg/kg
481（i）	硬脂酰乳酸钠	10 000 mg/kg
482（i）	硬脂酰乳酸钙	10 000 mg/kg
491	山梨醇酐单硬脂酸酯	5 000 mg/kg
492	山梨醇酐三硬脂酸酯	
493	山梨醇酐单月桂酸酯	
494	山梨醇酐单油酸酯	
495	山梨醇酐单棕榈酸酯	5 000 mg/kg
900a	聚二甲基硅氧烷	50 mg/kg
增味剂		
580	葡萄糖酸镁	根据GMP限量使用
620	$L(+)$-谷氨酸	
621	L-谷氨酸钠	
622	L-谷氨酸钾	
623	D, L-谷氨酸钙	
624	L-谷氨酸铵	
625	D, L-谷氨酸镁	
626	5′-鸟苷酸	
627	5′-鸟苷酸二钠	
628	5′-鸟苷酸二钾	
629	5′-鸟苷酸钙	
630	5′-肌苷酸	
631	5′-肌苷酸二钠	
632	5′-肌苷酸二钾	
633	5′-肌苷酸钙	

INS编号	添加剂名称	最大限量
634	5′-核糖核苷酸钙	根据GMP限量使用
635	5′-核糖核苷酸二钠	
636	麦芽酚	
637	乙基麦芽酚	
防腐剂		
200	山梨酸	1 000 mg/kg，以山梨酸计
202	山梨酸钾	
203	山梨酸钙	
210	苯甲酸	300 mg/kg，以苯甲酸计
211	苯甲酸钠	
212	苯甲酸钾	
213	苯甲酸钙	
234	乳酸链球菌素	500 mg/kg
稳定剂和增稠剂		
170（i）	碳酸钙	根据GMP限量使用
331（iii）	柠檬酸钠	
338	磷酸	1 000 mg/kg，单用或混用，以磷计
339（i）	磷酸二氢钠	
339（ii）	磷酸氢二钠	
339（iii）	磷酸三钠	
340（i）	磷酸二氢钾	
340（ii）	磷酸氢二钾	
340（iii）	磷酸三钾	
341（i）	磷酸二氢钙	
341（ii）	磷酸氢二钙	
341（iii）	磷酸三钙	
342（i）	磷酸二氢铵	
342（ii）	磷酸氢二铵	
343（i）	磷酸二氢镁	

INS编号	添加剂名称	最大限量
343（ii）	磷酸氢二镁	1 000 mg/kg，单用或混用，以磷计
343（iii）	磷酸三镁	
450（i）	磷酸二钠	
450（ii）	焦磷酸三钠	
450（iii）	焦磷酸四钠	
450（v）	二磷酸四钾	
450（vi）	磷酸二钙	
450（vii）	二磷酸二氢钙	
451（i）	三磷酸五钠	
451（ii）	三磷酸五钾	
452（i）	六偏磷酸钠	
452（ii）	聚磷酸钾	
452（iii）	聚磷酸钠钙	
452（iv）	聚磷酸钙	
452（v）	聚磷酸铵	
542	骨磷酸盐	
400	海藻酸	根据GMP限量使用
401	海藻酸钠	
402	海藻酸钾	
403	海藻酸铵	
404	海藻酸钙	
405	海藻酸丙二醇酯	
406	琼脂	
407	卡拉胶	
407a	加工琼芝属海藻胶（PES）	
410	槐豆胶（又名刺槐豆胶）	
412	瓜尔胶	
413	黄蓍胶	
414	阿拉伯胶	
415	黄原胶	

INS编号	添加剂名称	最大限量
416	刺梧桐胶	根据GMP限量使用
417	刺云实胶	
418	结冷胶	
425	魔芋粉	
440	果胶	
459	β-环状糊精	5 mg/kg
460（i）	微晶纤维素	根据GMP限量使用
460（ii）	纤维素粉	
461	甲基纤维素	
463	羟丙基纤维素	
464	羟丙基甲基纤维素	
465	甲基乙基纤维素	
466	羧甲基纤维素钠（纤维素胶）	
467	乙基羟乙基纤维素	
468	交联羧甲基纤维素钠（交联纤维素胶）	
469	酶解羧甲基纤维素钠（酶解纤维素胶）	
470（i）	肉豆蔻酸、棕榈酸和硬脂酸与氨、钙、钾和钠的盐	
470（ii）	油酸与钙、钾、钠的盐	
471	单、双甘油脂肪酸酯	
472a	乙酰化单、双甘油脂肪酸酯	
472b	乳酸脂肪酸甘油酯	
472c	柠檬酸脂肪酸甘油酯	
508	氯化钾	
509	氯化钙	
511	氯化镁	
1200	聚葡萄糖	
1400	糊精，焙炒淀粉	
1401	酸处理淀粉	
1402	碱处理淀粉	
1403	漂白淀粉	

INS编号	添加剂名称	最大限量
1404	氧化淀粉	根据GMP限量使用
1405	酶处理淀粉	
1410	单淀粉磷酸酯	
1412	磷酸酯双淀粉	
1413	磷酸化二淀粉磷酸酯	
1414	乙酰化二淀粉磷酸酯	
1420	醋酸酯淀粉	
1422	乙酰化双淀粉己二酸酯	
1440	羟丙基淀粉	
1442	羟丙基二淀粉磷酸酯	
1450	辛烯基琥珀酸淀粉钠	
1451	乙酰化氧化淀粉	
甜味剂(a)		
420	山梨醇	根据GMP限量使用
421	D-甘露糖醇	
950	安赛蜜	350 mg/kg
951	阿斯巴甜	1 000 mg/kg
952	甜蜜素	250 mg/kg
953	异麦芽酮糖醇（氢化异麦芽酮糖）	根据GMP限量使用
954	糖精钠	100 mg/kg
955	三氯蔗糖	400 mg/kg
956	阿力甜	100 mg/kg
961	纽甜	100 mg/kg
962	天门冬酰苯丙氨酸甲酯乙酰磺胺酸	350 mg/kg，以乙酰磺胺酸钾等量计
964	聚葡萄糖醇液	根据GMP限量使用
965	麦芽糖醇	
966	乳糖醇	
967	木糖醇	
968	赤藓糖醇	

（a）甜味剂仅限用于降低能量或不添加糖的乳或乳衍生产品。

5 污染物

本标准所涉及的产品应符合《食品及饲料中污染物和毒素通用标准》（CXS 193-1995）规定的污染物最大限量。

本标准所涉及的产品在加工中使用的牛乳应符合《食品及饲料中污染物和毒素通用标准》（CXS 193-1995）规定的乳中污染物和毒素最大限量以及国际食品法典委员会设定的农药和兽药最大残留限量。

6 卫生要求

建议本标准所涉及的产品应遵循《食品卫生总则》（CXC 1-1969）、《乳及乳制品卫生操作规范》（CXC 57-2004）以及《卫生操作规范》和《生产操作规范》等其他相关法典文本。产品应符合《食品微生物标准制定与实施原则和准则》（CXG 21-1997）规定的所有微生物标准。

7 标识

除符合《预包装食品标识通用标准》（CXS 1-1985）和《乳品术语》（CXS 206-1999）外，还应符合下列具体规定。

7.1 产品名称

7.1.1 第2.1条、第2.2条和第2.3条所涉及的产品名称可以根据情况适当命名为发酵乳或浓缩发酵乳。

如果该产品符合本标准的规定，可以用酸乳、嗜酸乳、开菲尔、马奶酒、Straisto、Labneh、Ymer和Ylette等替代这些名称。酸乳（Youghurt）可以由产品销售国决定其拼写。

对第2条所述"混合发酵酸乳"的品名，应使用一个适当的形容词与"酸乳"一词相结合。选定形容词应以准确且不误导消费者的方式说明，在酸乳制作过程通过选用发酵剂中特定乳酸杆菌使酸乳性质发生改变。与只标注"酸乳"的产品相比，此类改变可包括发酵有机物、代谢物和/或产品感官品质的显著不同。描述感官品质差异的形容词可包括"淡味"或"浓味"等词汇。"混合发酵酸乳"不得用作品名。

上述特定术语可能与"冷冻"相连，前提是：(i) 受到冷冻的产品符合本标准要求；(ii) 在解冻后有相当数目特定发酵剂能复活；(iii) 名称指定为冷冻产品并在售出后仅供直接食用。

只要命名在产品销售国不会对产品特性和标识形成误导，其他发酵乳和浓缩发酵乳可以使用产品销售国立法中规定的其他品种名称或现有常用名。

7.1.2 发酵后经热处理的发酵乳产品应命名为"热处理发酵乳"。如果这个名称会误导消费者，该产品将根据产品销售国立法允许的名称命名。在不存在这种立法的国家或也无其他通用名称的国家，该产品名称将是"热处理发酵乳"。

7.1.3　风味发酵乳品名应该包括主要调味物质或添加调味剂的名称。

7.1.4　第2.4条所述产品品名应为发酵乳饮料或采用产品销售国国家法律允许的其他名称。特别是作为配料加入发酵乳的水应在成分列表[1]中标注，发酵乳比例（m/m）应在标签中清晰注明。若经调味，则品名应包括主要调味物质或添加调味剂名称。

7.1.5　仅添加营养性碳水化合物甜味剂的发酵乳，可以标注为"含糖_____"，空格处可填入"发酵乳"或其他在第7.1.1条和第7.1.4条规定的命名。若使用非营养性甜味剂替代部分或全部食糖，则需要在标签上产品名称附近标注"使用_____增甜"或"使用_____及糖增甜"，空格处填写人工甜味剂名称。

7.1.6　本标准所涉及名称可以用于产品名称、标签、商业文件及广告设计，前提是将其用作配料，且配料特性在一定程度上得以保持，不会误导消费者。

7.2　乳脂含量说明

如果省略乳脂含量可能误导消费者，则应以产品最终消费国可以接受的方式声明：（i）质量百分比或体积百分比形式；（ii）如果产品标识上标明了份数，也可用每份中乳脂重量（g）表示。

7.3　非零售包装标识

在包装容器上除要标明产品名称、批次、生产厂家或包装商名称和地址之外，在包装容器上或附随说明书中，也应按本标准第7条和《预包装食品标识通用标准》（CXS 1-1985）第4.1~4.8条所要求的信息以及储存说明（在必要的地方）加以陈述。如果批次、生产厂家或包装商的名称和地址可以在附随的文件上标明，则可以用一个识别标识代替。

8　抽样和分析方法

为检查产品是否符合本标准，应采用《分析和抽样推荐性方法》（CXS 234-1999）中涉及本标准规定的分析和抽样方法。

1　见《预包装食品标签通用标准》（CXS 1-1985）第 4.2.1.5 条规定。

STANDARD FOR FERMENTED MILKS

CXS 243-2003

Adopted in 2003. Revised in 2008, 2010, 2018.

1 SCOPE

This standard applies to fermented milks, that is Fermented Milk including, Heat Treated Fermented Milks, Concentrated Fermented Milks and composite milk products based on these products, for direct consumption or further processing in conformity with the definitions in Section 2 of this Standard.

2 DESCRIPTION

2.1 *Fermented Milk* is a milk product obtained by fermentation of milk, which milk may have been manufactured from products obtained from milk with or without compositional modification as limited by the provision in Section 3.3, by the action of suitable microorganisms and resulting in reduction of pH with or without coagulation (iso-electric precipitation). These starter microorganisms shall be viable, active and abundant in the product to the date of minimum durability. If the product is heat treated after fermentation the requirement for viable microorganisms does not apply.

Certain Fermented Milks are characterized by specific starter culture(s) used for fermentation as follows:

Yoghurt:	Symbiotic cultures of *Streptococcus thermophilus* and *Lactobacillus delbrueckii* subsp. *bulgaricus*.
Alternate culture yoghurt:	Cultures of *Streptococcus thermophilus* and any *Lactobacillus* species.
Acidophilus milk:	*Lactobacillus acidophilus*.
Kefir:	Starter culture prepared from kefir grains, *Lactobacillus kefir*, species of the genera *Leuconostoc*, *Lactococcus* and *Acetobacter* growing in a strong specific relationship.
	Kefir grains constitute both lactose fermenting yeasts *(Kluyveromyces marxianus)* and non-lactose-fermenting yeasts *(Saccharomyces unisporus, Saccharomyces cerevisiae* and *Saccharomyces exiguus)*.
Kumys:	*Lactobacillus delbrueckii* subsp. *bulgaricus* and *Kluyveromyces marxianus*.

Other microorganisms than those constituting the specific starter culture(s) specified above may be added.

2.2 **Concentrated Fermented Milk** is a Fermented Milk the protein of which has been increased prior to or after fermentation to minimum 5.6%. Concentrated Fermented Milks includes traditional products such as Stragis to (strained yoghurt), Labneh, Ymer and Ylette.

2.3 **Flavoured Fermented Milks** are composite milk products, as defined in Section 2.3 of the *General Standard for the Use of Dairy Terms* (CXS 206-1999) which contain a maximum of 50% (m/m) of non-dairy ingredients (such as nutritive and non nutritive sweeteners, fruits and vegetables as well as juices, purees, pulps, preparations and preserves derived there from, cereals, honey, chocolate, nuts, coffee, spices and other harmless natural flavouring foods) and/or flavours. The non-dairy ingredients can be mixed in prior to/or after fermentation.

2.4 **Drinks based on Fermented Milk** are composite milk products, as defined in Section 2.3 of the *General Standard for the Use of Dairy Terms* (CXS 206-1999), obtained by mixing Fermented Milk as described in Section 2.1 with potable water with or without the addition of other ingredients such as whey, other non-dairy ingredients, and flavourings. Drinks Based on Fermented Milk contain a minimum of 40% (m/m) fermented milk.

Other microorganisms than those constituting the specific starter culture(s) specified above may be added.

3 ESSENTIAL COMPOSITION AND QUALITY FACTORS

3.1 Raw materials

- Milk and/or products obtained from milk;
- Potable water for the use in reconstitution or recombination.

3.2 Permitted ingredients

- Starter cultures of harmless microorganisms including those specified in Section 2;
- Other suitable and harmless microorganisms (*in products covered by Section 2.4*);
- Sodium chloride;
- Non-dairy ingredients as listed in Section 2.3 (Flavoured Fermented Milks);
- Potable water (*in products covered by Section 2.4*);
- Milk and milk products (*in products covered by Section 2.4*);
- Gelatine and starch in:
  - fermented milks heat-treated after fermentation;
  - flavoured fermented milk;
  - drinks based on fermented milk; and
  - plain fermented milks if permitted by national legislation in the country of sale to the final consumer.

provided they are added only in amounts functionally necessary as governed by Good Manufacturing Practice, taking into account any use of the stabilizers/thickeners listed in Section 4. These substances may be added either before or after adding the non-dairy ingredients.

3.3 Composition

	Fermented Milk	Yoghurt, Alternate Culture Yoghurt and Acidophilus milk	Kefir	Kumys
Milk protein[a] (% m/m)	min. 2.7%	min. 2.7%	min. 2.7%	
Milk fat (% m/m)	less than 10%	less than 15%	less than 10%	less than 10%
Titrable acidity, expressed as % lactic acid (% m/m)	min. 0.3%	min. 0.6%	min. 0.6%	min. 0.7%
Ethanol (% vol./w)				min. 0.5%
Sum of microorganisms constituting the starter culture defined in section 2.1 (cfu/g, in total)	min. 10^7	min. 10^7	min. 10^7	min. 10^7
Labelled microorganisms[b] (cfu/g, total)	min. 10^6	min. 10^6		
Yeasts (cfu/g)			min. 10^4	min. 10^4

(a) Protein content is 6.38 multiplied by the total Kjeldahl nitrogen determined.

(b) Applies where a content claim is made in the labelling that refers to the presence of a specific microorganism (other than those specified in section 2.1 for the product concerned) that has been added as a supplement to the specific starter culture.

In Flavoured Fermented Milks and Drinks based on Fermented Milk the above criteria apply to the fermented milk part. The microbiological criteria (based on the proportion of fermented milk product) are valid up to the date of minimum durability. This requirement does not apply to products heat-treated after fermentation.

Compliance with the microbiological criteria specified above is to be verified through analytical testing of the product through to "the date of minimum durability" after the product has been stored under the storage conditions specified in the labelling.

3.4 Essential manufacturing characteristics

Whey removal after fermentation is not permitted in the manufacture of fermented milks, except for Concentrated Fermented Milk (Section 2.2).

4 FOOD ADDITIVES

Only those additives classes indicated in the table below may be used for the product categories specified. Within each additive class, and where permitted according to the table, only those individual additives listed may be used and only within the limits specified.

In accordance with Section 4.1 of the Preamble to the *General Standard for Food Additives* (CXS 192-1995), additional additives may be present in the flavoured fermented milks and drinks based on fermented milk as a result of carry-over from non-dairy ingredients.

Additive class	Fermented Milks and Drinks based on Fermented Milk		Fermented Milks Heat Treated After Fermentation and Drinks based on Fermented Milk Heat Treated After Fermentation	
	Plain	Flavoured	Plain	Flavoured
Acidity regulators:	–	×	×	×
Carbonating agents:	×(b)	×(b)	×(b)	×(b)
Colours:	–	×	–	×
Emulsifiers:	–	×	–	×
Flavour enhancers:	–	×	–	×
Packaging gases:	–	×	×	×
Preservatives:	–	–	–	×
Stabilizers:	×(a)	×	×	×
Sweeteners:	–	×	–	×
Thickeners:	×(a)	×	×	×

(a) Use is restricted to reconstitution and recombination and if permitted by national legislation in the country of sale to the final consumer.
(b) Use of carbonating agents is technologically justified in Drinks based on Fermented Milk only.
× The use of additives belonging to the class is technologically justified. In the case of flavoured products the additives are technologically justified in the dairy portion.
– The use of additives belonging to the class is not technologically justified.

Acidity regulators, colours, emulsifiers, packaging gases and preservatives listed in Table3 of the *General Standard for Food Additives* (CXS 192-1995) are acceptable for use in fermented milk products categories as specified in the table above.

INS no.	Name of additive	Maximum level
Acidity regulators		
334	Tartaric acid *L*(+)-	2 000 mg/kg as tartaric acid
335(ii)	Sodium *L*(+)-tartrate	
337	Potassium sodium *L*(+)- tartrate	
355	Adipic acid	1 500 mg/kg as adipic acid
356	Sodium adipate	
357	Potassium adipate	
359	Ammonium adipate	

STANDARD FOR FERMENTED MILKS

INS no.	Name of additive	Maximum level
Carbonating agents		
290	Carbon dioxide	Limited by GMP
Colours		
100(i)	Curcumin	100 mg/kg
101(i)	Riboflavin, synthetic	300 mg/kg
101(ii)	Riboflavin 5'-phosphate, sodium	
102	Tartrazine	
104	Quinoline yellow	150 mg/kg
110	Sunset yellow FCF	300 mg/kg
120	Carmines	150 mg/kg
122	Azorubine (Carmoisine)	
124	Ponceau 4R (Cochineal red A)	
129	Allura red AC	300 mg/kg
132	Indigotine	100 mg/kg
133	Brilliant blue FCF	150 mg/kg
141(i)	Chlorophylls, copper complexes	500 mg/kg
141(ii)	Chlorophylls, copper complexes, sodium and potassium salts	
143	Fast green FCF	100 mg/kg
150b	Caramel II – sulphite caramel	150 mg/kg
150c	Caramel III – ammonia caramel	2 000 mg/kg
150d	Caramel IV – sulphite ammonia caramel	2 000 mg/kg
151	Brilliant black (Black PN)	150 mg/kg
155	Brown HT	150 mg/kg
160a(i)	Carotene, *beta*-, synthetic	100 mg/kg
160e	Carotenal, *beta*-apo-8'-	
160f	Carotenic acid, methyl or ethyl ester, *beta*-apo-8'-	
160a(iii)	Carotenes, *beta*-, *Blakeslea trispora*	

INS no.	Name of additive	Maximum level
160a(ii)	Carotenes, *beta*-, vegetable	600 mg/kg
160b(i)	Annatto extract, bixin- based	20 mg/kg as bixin
160b(ii)	Annatto extract, norbixin- based	20 mg/kg as norbixin
160d	Lycopenes	30 mg/kg as pure lycopene
161b(i)	Lutein from *Tagetes erecta*	150 mg/kg
161h(i)	Zeaxanthin, synthetic	150 mg/kg
163(ii)	Grape skin extract	100 mg/kg
172(i)	Iron oxide, black	100 mg/kg
172(ii)	Iron oxide, red	100 mg/kg
172(iii)	Iron oxide, yellow	100 mg/kg
Emulsifiers		
432	Polyoxyethylene (20) sorbitan monolaurate	3 000 mg/kg
433	Polyoxyethylene (20) sorbitan monooleate	3 000 mg/kg
434	Polyoxyethylene (20) sorbitan monopalmitate	3 000 mg/kg
435	Polyoxyethylene (20) sorbitan monostearate	3 000 mg/kg
436	Polyoxyethylene (20) sorbitan tristearate	3 000 mg/kg
472e	Diacetyltartaric and fatty acid esters of glycerol	10 000 mg/kg
473	Sucrose esters of fatty acids	5 000 mg/kg
474	Sucroglycerides	5 000 mg/kg
475	Polyglycerol esters of fatty acids	2 000 mg/kg
477	Propylene glycol esters of fatty acids	5 000 mg/kg
481(i)	Sodium stearoyl lactylate	10 000 mg/kg
482(i)	Calcium stearoyl lactylate	10 000 mg/kg
491	Sorbitan monostearate	5 000 mg/kg
492	Sorbitan tristearate	5 000 mg/kg
493	Sorbitan monolaurate	5 000 mg/kg
494	Sorbitan monooleate	5 000 mg/kg
495	Sorbitan monopalmitate	5 000 mg/kg

INS no.	Name of additive	Maximum level
900a	Polydimethylsiloxane	50 mg/kg
Flavour enhancers		
580	Magnesium gluconate	Limited by GMP
620	Glutamic acid, (*L*+)-	
621	Monosodium *L*-glutamate	
622	Monopotassium *L*-glutamate	
623	Calcium di-*L*-glutamate	
624	Monoammonium *L*-glutamate	
625	Magnesium di-*L*-glutamate	
626	Guanylic acid, 5'-	
627	Disodium 5'-guanylate-	
628	Dipotassium 5'-guanylate-	
629	Calcium 5'-guanylate	
630	Inosinic acid, 5'-	
631	Disodium 5'-inosinate	
632	Dipotassium 5'-inosinate	
633	Calcium 5'-inosinate	
634	Calcium 5'-ribonucleotides-	
635	Disodium 5'-ribonucleotides-	
636	Maltol	
637	Ethyl maltol	
Preservatives		
200	Sorbic acid	1 000 mg/kg as sorbic acid
202	Potassium sorbate	
203	Calcium sorbate	
210	Benzoic acid	300 mg/kg as benzoic acid
211	Sodium benzoate	
212	Potassium benzoate	
213	Calcium benzoate	300 mg/kg as benzoic acid

INS no.	Name of additive	Maximum level
234	Nisin	500 mg/kg
Stabilizers and Thickeners		
170(i)	Calcium carbonate	Limited by GMP
331(iii)	Trisodium citrate	
338	Phosphoric acid	1 000 mg/kg, singly or in combination, as phosphorus
339(i)	Sodium dihydrogen phosphate	
339(ii)	Disodium hydrogen phosphate	
339(iii)	Trisodium phosphate	
340(i)	Potassium dihydrogen phosphate	
340(ii)	Dipotassium hydrogen phosphate	
340(iii)	Tripotassium phosphate	
341(i)	Monocalcium dihydrogen phosphate	
341(ii)	Calcium hydrogen phosphate	
341(iii)	Tricalcium orthophosphate	
342(i)	Ammonium dihydrogen phosphate	
342(ii)	Diammonium hydrogen phosphate	
343(i)	Monomagnesium phosphate	
343(ii)	Magnesium hydrogen phosphate	
343(iii)	Trimagnesium phosphate	
450(i)	Disodium diphosphate	
450(ii)	Trisodium diphosphate	
450(iii)	Tetrasodium diphosphate	
450(v)	Tetrapotassium diphosphate	
450(vi)	Dicalcium diphosphate	
450(vii)	Calcium dihydrogen diphosphate	
451(i)	Pentasodium triphosphate	
451(ii)	Pentapotassium triphosphate	
452(i)	Sodium polyphosphate	

INS no.	Name of additive	Maximum level
452(ii)	Potassium polyphosphate	1 000 mg/kg, singly or in combination, as phosphorus
452(iii)	Sodium calcium polyphosphate	
452(iv)	Calcium polyphosphate	
452(v)	Ammonium polyphosphate	
542	Bone phosphate	
400	Alginic acid	Limited by GMP
401	Sodium alginate	
402	Potassium alginate	
403	Ammonium alginate	
404	Calcium alginate	
405	Propylene glycol alginate	
406	Agar	
407	Carrageenan	
407a	Processed euchema seaweed (PES)	
410	Carob bean gum	
412	Guar gum	
413	Tragacanth gum	
414	Gum Arabic (Acacia gum)	
415	Xanthan gum	
416	Karaya gum	
417	Tara gum	
418	Gellan gum	
425	Konjac flour	
440	Pectins	
459	Cyclodextrin, -beta	5 mg/kg
460(i)	Microcrystalline cellulose (Cellulose gel)	Limited by GMP
460(ii)	Powdered cellulose	
461	Methyl cellulose	
463	Hydroxypropyl cellulose	

INS no.	Name of additive	Maximum level
464	Hydroxypropyl methyl cellulose	Limited by GMP
465	Methyl ethyl cellulose	
466	Sodium carboxymethyl cellulose (Cellulose gum)	
467	Ethyl hydroxyethyl cellulose	
468	Cross-linked sodium carboxymethylcellulose (Cross-linked cellulose gum)	
469	Sodium carboxymethyl cellulose, enzymatically hydrolyzed (Cellulosegum, enzymatically hydrolyzed)	
470(i)	Salts of myristic, palmitic and stearic acids with ammonia, calcium, potassium and sodium	
470(ii)	Salts of oleic acid with calcium, potassium and sodium	
471	Mono- and di- glycerides of fatty acids	
472a	Acetic and fatty acid esters of glycerol	
472b	Lactic and fatty acid esters of glycerol	
472c	Citric and fatty acid esters of glycerol	
508	Potassium chloride	
509	Calcium chloride	
511	Magnesium chloride	
1200	Polydextrose	
1400	Dextrins, roasted starch	
1401	Acid treated starch	
1402	Alkaline treated starch	
1403	Bleached starch	
1404	Oxidized starch	
1405	Starches, enzyme treated	
1410	Mono starch phosphate	
1412	Distarch phosphate	
1413	Phosphated distarch phosphate	

INS no.	Name of additive	Maximum level
1414	Acetylated distarch phosphate	Limited by GMP
1420	Starch acetate	
1422	Acetylated distarch adipate	
1440	Hydroxypropyl starch	
1442	Hydroxypropyl distarch phosphate	
1450	Starch sodium octenyl succinate	
1451	Acetylated oxidized starch	
Sweeteners[a]		
420	Sorbitol	Limited by GMP
421	Mannitol	
950	Acesulfame potassium	350 mg/kg
951	Aspartame	1 000 mg/kg
952	Cyclamates	250 mg/kg
953	Isomalt (Hydrogenated isomaltulose)	Limited by GMP
954	Saccharin	100 mg/kg
955	Sucralose (Trichlorogalactosucrose)	400 mg/kg
956	Alitame	100 mg/kg
961	Neotame	100 mg/kg
962	Aspartame-acesulfame salt	350 mg/kg, on an acesulfame potassium equivalent basis
964	Polyglycitol syrup	Limited by GMP
965	Maltitols	
966	Lactitol	
967	Xylitol	
968	Erythritol	

(a) The use of sweeteners is limited to milk-and milk derivative-based products energy reduced or with no added sugar

5 CONTAMINANTS

The products covered by this Standard shall comply with the Maximum Levels for contaminants that are specified for the product in the *General Standard for Contaminants and Toxins in Foods and Feeds* (CXS 193-1995).

The milk used in the manufacture of the products covered by this Standard shall comply with the Maximum Levels for contaminants and toxins specified for milk by the *General Standard for Contaminants and Toxins in Foods and Feeds* (CXS 193-1995) and with the maximum residue limits for veterinary drug residues and pesticides established for milk by the CAC.

6   HYGIENE

It is recommended that the products covered by the provisions of this standard be prepared and handled in accordance with the appropriate sections of the *General Principles of Food Hygiene* (CXC 1-1969), the *Code of Hygienic Practice for Milk and Milk Products* (CXC 57-2004) and other relevant Codex texts such as Codes of Hygienic Practice and Codes of Practice. The products should comply with any microbiological criteria established in accordance with the *Principles and Guidelines for the Establishment and Application of Microbiological Criteria Related to Foods* (CXG 21-1997).

7   LABELLING

In addition to the provisions of the *General Standard for the Labelling of Prepackaged Foods* (CXS 1-1985) and the *General Standard for the Use of Dairy Terms* (CXS 206-1999), the following specific provisions apply:

7.1   Name of the food

7.1.1   The name of the products covered by Sections 2.1, 2.2 and 2.3, shall be fermented milk or concentrated fermented milk as appropriate.

However, these names may be replaced by the designations Yoghurt, Acidophilus Milk, Kefir, Kumys, Stragisto, Labneh, Ymer and Ylette, provided that the product complies with the specific provisions of this Standard. Yoghurt may be spelled as appropriate in the country of retail sale.

"Alternate culture yoghurt", as defined in Section 2, shall be named through the use of an appropriate qualifier in conjunction with the word "yoghurt". The chosen qualifier shall describe, in a way that is accurate and not misleading to the consumer, the nature of the change imparted to the yoghurt through the selection of the specific *Lactobacilli* in the culture for manufacturing the product. Such change may include a marked difference in the fermentation organisms, metabolites and/or sensory properties of the product when compared to the product designated solely as "yoghurt". Examples of qualifiers which describe differences in sensory properties include terms such as "mild" and "tangy". The term "alternate culture yoghurt" shall not apply as a designation.

The above specific terms may be used in connection with the term "frozen" provided(i) that the product submitted to freezing complies with the requirements in this Standard, (ii) that the specific starter cultures can be reactivated in reasonable numbers by thawing, and (iii) that the frozen product is named as such and is sold for direct consumption, only.

Other fermented milks and concentrated fermented milks may be designated with other variety names as specified in the national legislation of the country in which the product is sold, or names existing by common usage, provided that such designations do not create an erroneous impression in the country of retail sale regarding the character and identity of the food.

7.1.2 Products obtained from fermented milk(s) heat treated after fermentation shall be named "Heat Treated Fermented Milk". If the consumer would be misled by this name, the products shall be named as permitted by national legislation in the country of retail sale. In countries where no such legislation exists, or no other names are in common usage, the product shall be named "Heat Treated Fermented Milk".

7.1.3 The designation of Flavoured Fermented Milks shall include the name of the principal flavouring substance(s) or flavour(s) added.

7.1.4 The name of the products defined in Section 2.4 shall be drinks based on fermented milk or may be designated with other variety names as allowed in the national legislation of the country in which the product is sold. In particular, water added as an ingredient to fermented milk shall be declared in the list of ingredients[1] and the percentage of fermented milk used (m/m) shall clearly appear on the label. When flavoured, the designation shall include the name of the principal flavouring substance(s) or flavour(s) added.

7.1.5 Fermented milks to which only nutritive carbohydrate sweeteners have been added, maybe labelled as "sweetened _____", the blank being replaced by the term "Fermented Milk" or another designation as specified in Section 7.1.1 and 7.1.4. If non-nutritive sweeteners are added in partial or total substitution to sugar, the mention "sweetened with _____" or "sugared and sweetened with _____" should appear close to the name of the product, the blank being filled in with the name of the artificial sweeteners.

7.1.6 The names covered by this Standard may be used in the designation, on the label, in commercial documents and advertising of other foods, provided that it is used as an ingredient and that the characteristics of the ingredient are maintained to a relevant degree in order not to mislead the consumer.

7.2 **Declaration of fat content**

If the consumer would be misled by the omission, the milk fat content shall be declared in a manner acceptable in the country of sale to the final consumer, either as(i) a percentage of mass or volume, or (ii) in grams per serving as qualified in the label, provided that the number of servings is stated.

7.3 **Labelling of non-retail containers**

Information required in Section 7 of this Standard and Sections 4.1 to 4.8 of the *General Standard for the Labelling of Pre-packaged Foods*, and, if necessary, storage instructions, shall be given either on the container or in accompanying documents, except that the name of the product, lot identification, and the name and address of the manufacturer or packer, shall appear on the container. However, lot identification and the name and address of the manufacturer or packager may be replaced by an identification mark, provided that such mark is clearly identifiable with the accompanying documents.

8 **METHODS OF SAMPLING AND ANALYSIS**

For checking the compliance with this standard, the methods of analysis and sampling contained in the *Recommended Methods of Analysis and Sampling* (CXS 234-1999) relevant to the provisions in this standard, shall be used.

---

1 As prescribed in section 4.2.1.5 of the *General Standard for the Labelling of Prepackaged Foods* (CXS 1-1985).

脱脂乳与植物脂肪混合物

STANDARD FOR A BLEND OF EVAPORATED SKIMMED MILK AND VEGETABLE FAT

CXS 250-2006

2006年通过。2010年、2018年修订。

1 范围

本标准适用于符合本标准的第2条所述即食或用于进一步加工的脱脂乳与植物脂肪混合物，也称无糖脱脂乳与植物脂肪的混合浓缩物。

2 说明

脱脂乳与植物脂肪混合物是指重组乳成分和饮用水，或进行部分脱水并添加食用植物油、食用植物脂肪及其混合物，满足本标准第3条的成分要求。

3 基本成分和质量要求

3.1 基本原料

脱脂乳和脱脂乳粉[1]，其他非脂乳固体和可食的植物脂肪/油[1]。

允许用下列乳制品调整蛋白质含量：

- 乳超滤滞留物：乳超滤滞留物是指乳、部分脱脂乳或脱脂乳，通过超滤法浓缩乳蛋白后形成的产品；
- 乳渗透物：乳渗透物是指通过超滤法，将乳蛋白和乳脂从乳、部分脱脂乳或脱脂乳中分离除去而形成的产品；
- 乳糖[1]。

3.2 其他配料

- 饮用水；
- 氯化钠和/或氯化钾作为替代盐。

3.3 允许营养素

根据《食品中必需营养素添加通则》（CXG 9-1987），由各个国家立法，按需要酌情制定维生素A、维生素D和其他营养素的最高和最低含量，包括禁止使用某些特定营养素。

---

[1] 规格参见相关国际食品法典标准。

3.4 成分

脱脂乳与植物脂肪混合物

总脂肪最低含量	7.5% m/m
非脂乳固体最低含量[a]	17.5% m/m
乳的非脂固体中乳蛋白最低含量[a]	34% m/m

低脂的脱脂乳与植物脂肪混合物

总脂肪含量	>1%，<7.5% m/m
非脂乳固体含量[a]	19% m/m
乳的非脂固体中乳蛋白最低含量[a]	34% m/m

（a）乳的非脂固体包括乳糖的结晶水。

4 食品添加剂

只允许在规定范围内使用下表列出的食品添加剂。

INS编号	添加剂名称	最大限量
乳化剂		
322	卵磷脂	根据GMP限量使用
稳定剂		
331（i）	柠檬酸二氢钠	根据GMP限量使用
331（iii）	柠檬酸钠	
332（i）	柠檬酸二氢钾	
332（ii）	柠檬酸钾	
333	柠檬酸钙	
508	氯化钾	根据GMP限量使用
509	氯化钙	
酸度调节剂		
170（i）	碳酸钙	根据GMP限量使用
339（i）	磷酸二氢钠	4 400 mg/kg，单用或混用，以磷计
339（ii）	磷酸氢二钠	
339（iii）	磷酸三钠	
340（i）	磷酸二氢钾	
340（ii）	磷酸氢二钾	
340（iii）	磷酸三钾	

INS编号	添加剂名称	最大限量
341（i）	磷酸二氢钙	4 400 mg/kg，单用或混用，以磷计
341（ii）	磷酸氢二钙	
341（iii）	磷酸三钙	
450（i）	磷酸二钠	
450（ii）	焦磷酸三钠	
450（iii）	焦磷酸四钠	
450（v）	二磷酸四钾	
450（vi）	磷酸二钙	
450（vii）	二磷酸二氢钙	
451（i）	三磷酸五钠	
451（ii）	三磷酸五钾	
452（i）	六偏磷酸钠	
452（ii）	聚磷酸钾	
452（iii）	聚磷酸钠钙	
452（iv）	聚磷酸钙	
452（v）	聚磷酸铵	
500（i）	碳酸钠	根据GMP限量使用
500（ii）	碳酸氢钠	
500（iii）	倍半碳酸钠	
501（i）	碳酸钾	
501（ii）	碳酸氢钾	
增稠剂		
407	卡拉胶	根据GMP限量使用
407a	加工琼芝属海藻胶（PES）	

5　污染物

本标准所涉及的产品应符合《食品及饲料中污染物和毒素通用标准》（CXS 193-1995）规定的污染物最大限量。

本标准所涉及的产品在加工中使用的牛乳应符合《食品及饲料中污染物和毒素通用标准》（CXS 193-1995）规定的乳中污染物和毒素最大限量以及国际食品法典委员会设定的农药和兽药最大残留限量。

本标准所涉及的产品生产所用植物油/脂肪应遵循《食品及饲料中污染物和毒素通用标准》（CXS 193-1995）中规定的油/脂肪中污染物和毒素最大限量以及国际食品法典委员会设定的油/脂肪中农药最大残留限量。

6 卫生要求

建议本标准所涉及的产品应遵循《食品卫生总则》（CXC 1-1969）、《乳及乳制品卫生操作规范》（CXC 57-2004）以及《卫生操作规范》和《生产操作规范》等其他相关法典文本。产品应符合根据《食品微生物标准制定与实施原则和准则》（CXG 21-1997）规定的所有微生物标准。

7 标识

除符合《预包装食品标识通用标准》（CXS 1-1985）外，还应符合下列具体规定。

7.1 产品名称

产品名称应为：

— 脱脂乳与植物脂肪混合物；
— 低脂的脱脂乳和植物脂肪混合物。

如果产品销售国立法允许，也可用其他名称。

7.2 乳脂含量说明

总脂肪含量应以销售国可接受的方式向最终消费者声明：（i）质量百分比或体积百分比形式；（ii）如果产品标识上标明了份数，也可用每份中乳脂重量（g）表示。标识上需说明是否含有可食用植物脂肪或可食用植物油。当销售国有要求时，脂肪或油的植物来源应包含在食品名称中或另附说明。

7.3 乳蛋白含量说明

乳蛋白含量应以销售国可接受的方式向最终消费者声明：（i）质量百分比或体积百分比形式；（ii）如果产品标识上标明了份数，也可用每份中乳蛋白重量（g）表示。

7.4 配料表

尽管《预包装食品标识通用标准》（CXS 1-1985）中第4.2.1条有规定，但仅用于调节乳蛋白含量的乳制品不需要声明。

7.5 建议性说明

标识上要有指示该产品不能代替婴儿配方食品的说明。例如"不适用婴儿"。

8 抽样和分析方法

为检查产品是否符合本标准，应采用《分析和抽样推荐性方法》（CXS 234-1999）中涉及本标准规定的分析和抽样方法。

STANDARD FOR A BLEND OF EVAPORATED SKIMMED MILK AND VEGETABLE FAT

CXS 250-2006

Adopted in 2006. Amended in 2010, 2018.

1   SCOPE

This Standard applies to a blend of evaporated skimmed milk and vegetable fat, also known as a blend of unsweetened condensed skimmed milk and vegetable fat, which is intended for direct consumption, or further processing, in conformity with the description in Section 2 of this Standard.

2   DESCRIPTION

A blend of evaporated skimmed milk and vegetable fat is a product prepared by recombining milk constituents and potable water, or by the partial removal of water and the addition of edible vegetable oil, edible vegetable fat or a mixture thereof, to meet the compositional requirements in Section 3 of this Standard.

3   ESSENTIAL COMPOSITION AND QUALITY FACTORS

3.1   Raw materials

Skimmed milk and skimmed milk powders[1], other non-fat milk solids, and edible vegetable fats/oils[1].

The following milk products are allowed for protein adjustment purposes:

- Milk retentate: Milk retentate is the product obtained by concentrating milk protein by ultra-filtration of milk, partly skimmed milk, or skimmed milk;
- Milk permeate: Milk permeate is the product obtained by removing milk protein and milk fat from milk, partly skimmed milk, or skimmed milk by ultra-filtration; and
- Lactose[1].

3.2   Permitted ingredients

- Potable water;
- Sodium chloride and/or potassium chloride as salt substitute.

3.3   Permitted nutrients

Where allowed in accordance with the *General Principles for the Addition of Essential Nutrients to Food* (CXG 9-1987), maximum and minimum levels for Vitamins A, D and other nutrients, where appropriate, should be laid down by national legislation in accordance with the needs of individual country including, where appropriate, the prohibition of the use of particular nutrients.

---

1   For specification, see relevant Codex Standard.

STANDARD FOR A BLEND OF EVAPORATED SKIMMED MILK AND VEGETABLE FAT

3.4 Composition

Blend of evaporated skimmed milk and vegetable fat

Minimum total fat	7.5% m/m
Minimum milk solids-not-fat[a]	17.5% m/m
Minimum milk protein in milk solids-not-fat[a]	34% m/m

Reduced fat blend of evaporated skimmed milk and vegetable fat

Total fat	More than 1% and less than 7.5% m/m
Minimum milk solids-not-fat[a]	19% m/m
Minimum milk protein in milk solids-not-fat[a]	34% m/m

(a) The milk solids-not-fat content includes water of crystallization of the lactose.

4 FOOD ADDITIVES

Only food additives listed below may be used and only within the limits specified.

INS no.	Name of additive	Maximum level
Emulsifiers		
322	Lecithins	Limited by GMP
Stabilizers		
331(i)	Sodium dihydrogen citrate	Limited by GMP
331(iii)	Trisodium citrate	
332(i)	Potassium dihydrogen citrate	
332(ii)	Tripotassium citrate	
333	Calcium citrate	
508	Potassium chloride	
509	Calcium chloride	
Acidity regulators		
170(i)	Calcium carbonate	Limited by GMP
339(i)	Sodium dihydrogen phosphate	4 400 mg/kg, singly or in combination as phosphorous
339(ii)	Disodium hydrogen phosphate	
339(iii)	Trisodium phosphate	
340(i)	Potassium dihydrogen phosphate	
340(ii)	Dipotassium hydrogenphosphate	
340(iii)	Tripotassium phosphate	
341(i)	Calcium dihydrogen phosphate	
341(ii)	Dicalcium hydrogen phosphate	
341(iii)	Tricalcium phosphate	

INS no.	Name of additive	Maximum level
450(i)	Disodium diphosphate	4 400 mg/kg, singly or in combination as phosphorous
450(ii)	Trisodium diphosphate	
450(iii)	Tetrasodium diphosphate	
450(v)	Tetrapotassium diphosphate	
450(vi)	Dicalcium diphosphate	
450(vii)	Calcium dihydrogen diphosphate	
451(i)	Pentasodium triphosphate	
451(ii)	Pentapotassium triphosphate	
452(i)	Sodium polyphosphate	
452(ii)	Potassium polyphosphate	
452(iii)	Sodium calcium polyphosphate	
452(iv)	Calcium polyphosphates	
452(v)	Ammonium polyphosphates	
500(i)	Sodium carbonate	Limited by GMP
500(ii)	Sodium hydrogen carbonate	
500(iii)	Sodium sesquicarbonate	
501(i)	Potassium carbonates	
501(ii)	Potassium hydrogen carbonate	
Thickeners		
407	Carrageenan	Limited by GMP
407a	Processed euchema seaweed (PES)	

5 CONTAMINANTS

The products covered by this Standard shall comply with the maximum levels for contaminants that are specified for the product in the *General Standard for Contaminants and Toxins in Food and Feed* (CXS 193-1995).

The milk used in the manufacture of the products covered by this Standard shall comply with the maximum levels for contaminants and toxins specified for milk by the *General Standard for Contaminants and Toxins in Food and Feed* (CXS 193-1995) and with the maximum residue limits for veterinary drug residues and pesticides established for milk by the CAC.

The vegetable oils/fat used in the manufacture of the products covered by this Standard shall comply with the Maximum Levels for contaminants and toxins specified for the oils/fats by the *General Standard for Contaminants and Toxins in Food and Feed* (CXS 193-1995) and with the maximum residue limits for pesticides established for the oils/fats by the CAC.

STANDARD FOR A BLEND OF EVAPORATED SKIMMED MILK AND VEGETABLE FAT

6 HYGIENE

It is recommended that the products covered by the provisions of this Standard be prepared and handled in accordance with the appropriate sections of the *General Principles of Food Hygiene* (CXC 1-1969), the *Code of Hygienic Practice for Milk and Milk Products* (CXC 57-2004) and other relevant Codex texts such as Codes of Hygienic Practice and Codes of Practice. The products should comply with any microbiological criteria established in accordance with the *Principles and Guidelines for the Establishment and Application of Microbiological Criteria Related to Foods* (CXG 21-1997).

7 LABELLING

In addition to the provision of the *General Standard for the Labelling of Prepackaged Foods* (CXS 1-1985) the following specific provisions apply.

7.1 Name of the food

The name of the food shall be:

- Blend of Evaporated Skimmed Milk and Vegetable Fat; or
- Reduced Fat Blend of Evaporated Skimmed Milk and Vegetable Fat.

Other names may be used if allowed by national legislation in the country of retail sale.

7.2 Declaration of total fat content

The total fat content shall be declared in a manner found acceptable in the country of sale to the final consumer, either (i) as a percentage by mass or volume, or (ii) in grams per serving as quantified in the label provided that the number of servings is stated.

A statement shall appear on the label as to the presence of edible vegetable fat and/or edible vegetable oil. When required by the country of retail sale, the common name of the vegetable from which the fat or oil is derived shall be included in the name of the food or as a separate statement.

7.3 Declaration of milk protein

The milk protein content shall be declared in a manner acceptable in the country of sale to the final consumer, either (i) as a percentage by mass or volume, or (ii) in grams per serving as quantified in the label provided that the number of servings is stated.

7.4 List of ingredients

Notwithstanding the provision of Section 4.2.1 of the *General Standard for the Labelling of Prepackaged Foods* (CXS 1-1985) milk products used only for protein adjustment need not be declared.

7.5 Advisory statement

A statement shall appear on the label to indicate that the product should not be used as a substitute for infant formula. For example, "NOT SUITABLE FOR INFANTS".

8 METHODS OF SAMPLING AND ANALYSIS

For checking the compliance with this Standard, the methods of analysis and sampling contained in the *Recommended Methods of Analysis and Sampling* (CXS 234-1999) relevant to the provisions in this Standard, shall be used.

脱脂乳与植物脂肪混合粉剂

STANDARD FOR A BLEND OF SKIMMED MILK AND VEGETABLE FAT IN POWDERED FORM

CXS 251-2006

2006年通过。2010年、2013年、2014年、2016年、2018年修订。

1 范围

本标准适用于符合本标准第2条所述即食或用于进一步加工的脱脂乳与植物脂肪混合粉剂。

2 说明

脱脂乳与植物脂肪混合粉剂是指进行部分脱水并添加食用植物油、食用植物脂肪及其混合物制成的产品，满足本标准第3条的成分要求。

3 基本成分和质量要求

3.1 基本原料

脱脂乳和脱脂乳粉[1]，其他非脂乳固体和可食植物脂肪/油[1]。

允许用下列乳制品调整蛋白质含量：

- 乳超滤滞留物：乳超滤滞留物是指乳、部分脱脂乳或脱脂乳，通过超滤法浓缩乳蛋白后形成的产品；
- 乳渗透物：乳渗透物是指通过超滤法，将乳蛋白和乳脂从乳、部分脱脂乳或脱脂乳中分离除去形成的产品；
- 乳糖[1]。

3.2 允许营养素

根据《食品中必需营养素添加通则》（CXG 9-1987），由各个国家立法，按需酌情制定维生素A、维生素D和其他营养素的最高和最低含量，包括禁止使用某些特定营养素。

3.3 成分

脱脂乳与植物脂肪混合粉剂

总脂肪最低含量　　　　　　　　　　26% m/m

---

[1] 规格参见相关国际食品法典标准。

水分最高含量[a]	5% m/m
乳的非脂固体中乳蛋白最低含量[a]	34% m/m

低脂的脱脂乳与植物脂肪混合粉剂

总脂肪含量	>1.5%，<26% m/m
水分最高含量[a]	5% m/m
乳的非脂固体中乳蛋白最低含量[a]	34% m/m

(a) 乳的非脂固体包括乳糖的结晶水。

4　食品添加剂

只允许在规定范围内使用下表列出的食品添加剂。

INS编号	添加剂名称	最大限量
稳定剂		
331（i）	柠檬酸二氢钠	根据GMP限量使用
331（iii）	柠檬酸钠	
332（i）	柠檬酸二氢钾	
332（ii）	柠檬酸钾	
508	氯化钾	
509	氯化钙	
酸度调节剂		
339（i）	磷酸二氢钠	4 400 mg/kg，单用或混用，以磷计
339（ii）	磷酸氢二钠	
339（iii）	磷酸三钠	
340（i）	磷酸二氢钾	
340（ii）	磷酸氢二钾	
340（iii）	磷酸三钾	
341（i）	磷酸二氢钙	
341（ii）	磷酸氢二钙	
450（i）	磷酸二钠	
450（ii）	焦磷酸三钠	
450（iii）	焦磷酸四钠	

INS编号	添加剂名称	最大限量
450（v）	二磷酸四钾	4 400 mg/kg，单用或混用，以磷计
450（vi）	磷酸二钙	
450（vii）	二磷酸二氢钙	
451（i）	三磷酸五钠	
451（ii）	三磷酸五钾	
452（i）	六偏磷酸钠	
452（ii）	聚磷酸钾	
452（iii）	聚磷酸钠钙	
452（iv）	聚磷酸钙	
452（v）	聚磷酸铵	
500（i）	碳酸钠	根据GMP限量使用
500（ii）	碳酸氢钠	
500（iii）	倍半碳酸钠	
501（i）	碳酸钾	
501（ii）	碳酸氢钾	
乳化剂		
322	卵磷脂	根据GMP限量使用
471	单、双甘油脂肪酸酯	
抗结块剂		
170（i）	碳酸钙	根据GMP限量使用
504（i）	碳酸镁	
530	氧化镁	
551	二氧化硅（无定型）	
552	硅酸钙	
553（i）	硅酸镁（合成）	
553（iii）	滑石粉	
554	硅酸铝钠	570 mg/kg，以铝计
341（iii）	磷酸三钙	4 400 mg/kg，单用或混用，以磷计
343（iii）	磷酸三镁	

INS编号	添加剂名称	最大限量
抗氧化剂		
300	L-抗坏血酸	500 mg/kg，以抗坏血酸计
301	抗坏血酸钠	
304	抗坏血酸棕榈酸酯	80 mg/kg，单用或混用，以抗坏血酸硬脂酸酯计
305	抗坏血酸硬脂酸酯	
319	特丁基对苯二酚	100 mg/kg，单用或混用，以脂肪或油表示
320	丁基羟基茴香醚	
321	丁基羟基甲苯	

5 污染物

本标准所涉及的产品应符合《食品及饲料中污染物和毒素通用标准》（CXS 193-1995）规定的污染物最大限量。

本标准所涉及的产品在加工中使用的牛乳应符合《食品及饲料中污染物和毒素通用标准》（CXS 193-1995）规定的乳中污染物和毒素最大限量以及国际食品法典委员会设定的农药和兽药最大残留限量。

本标准所涉及的产品在生产中所用植物油/脂肪应遵循《食品及饲料中污染物和毒素通用标准》（CXS 193-1995）中规定的油/脂肪中污染物和毒素最大限量以及国际食品法典委员会设定的油/脂肪中农药最大残留限量。

6 卫生要求

建议本标准所涉及的产品应遵循《食品卫生总则》（CXC 1-1969）、《乳及乳制品卫生操作规范》（CXC 57-2004）以及《卫生操作规范》和《生产操作规范》等其他相关法典文本。产品应符合《食品微生物标准制定与实施原则和准则》（CXG 21-1997）规定的所有微生物标准。

7 标识

除符合《预包装食品标识通用标准》（CXS 1-1985）外，还应符合下列具体规定。

7.1 产品名称

产品名称应为：

— 脱脂乳和植物脂肪混合粉剂；
— 低脂脱脂乳和植物脂肪混合粉剂。

如果产品销售国立法允许，也可以用其他名称。

7.2 乳脂含量说明

总脂肪含量应以销售国可接受的方式向最终消费者声明：（i）质量百分比或体积百分比形式；（ii）如果产品标识上标明了份数，也可用每份中乳脂重量（g）表示。标识上需说明是否含有可食用植物脂肪或可食用植物油。当销售国有要求时，脂肪或油的植物来源应包含在食品名称中或另附说明。

7.3 乳蛋白含量说明

乳蛋白含量应以销售国可接受的方式向最终消费者声明：（i）质量百分比或体积百分比形式；（ii）如果产品标识上标明了份数，也可用每份中乳蛋白重量（g）表示。

7.4 配料表

尽管《预包装食品标识通用标准》（CXS 1-1985）中第4.2.1条有规定，但仅用于调节乳蛋白含量的乳制品不需要声明。

7.5 建议性说明

标识上要有指示该产品不能代替婴儿配方食品的说明。例如"婴儿不适用"。

8 抽样和分析方法

为检查产品是否符合本标准，应采用《分析和抽样推荐性方法》（CXS 234-1999）中涉及本标准规定的分析和抽样方法。

STANDARD FOR A BLEND OF SKIMMED MILK AND VEGETABLE FAT IN POWDERED FORM

CXS 251-2006

Adopted in 2006. Amended in 2010, 2013, 2014, 2016, 2018.

1 SCOPE

This Standard applies to a blend of skimmed milk and vegetable fat in powdered form, intended for direct consumption, or further processing, in conformity with the description in Section 2 of this Standard.

2 DESCRIPTION

A blend of skimmed milk and vegetable fat in powdered form is a product prepared by the partial removal of water from milk constituents with the addition of edible vegetable oil, edible vegetable fat or a mixture thereof, to meet the compositional requirements in Section 3 of this Standard.

3 ESSENTIAL COMPOSITION AND QUALITY FACTORS

3.1 Raw materials

Skimmed milk and skimmed milk powders[1], other non-fat milk solids, and edible vegetable oils/ fats[1].

The following milk products are allowed for protein adjustment purposes:

- Milk retentate: Milk retentate is the product obtained by concentrating milk protein by ultrafiltration of milk, partly skimmed milk, or skimmed milk;
- Milk permeate: Milk permeate is the product obtained by removing milk proteins and milk fat from milk, partly skimmed milk or skimmed milk by ultrafiltration; and
- Lactose[1].

3.2 Permitted nutrients

Where allowed in accordance with the *General Principles for the Addition of Essential Nutrients for Food* (CXG 9-1987), maximum and minimum levels for Vitamins A, D and other nutrients, where appropriate, should be laid down by national legislation in accordance with the needs of individual country including, where appropriate, the prohibition of the use of particular nutrients.

3.3 Composition

Blend of skimmed milk and vegetable fat in powdered form

Minimum total fat	26% m/m
Maximum water[a]	5% m/m
Minimum milk protein in milk solids-not-fat[a]	34% m/m

---

[1] For specification, see relevant Codex Standard.

Reduced fat blend of skimmed milk powder and vegetable fat in powdered form

Total fat	More than 1.5% and less than 26% m/m
Maximum water[a]	5% m/m
Minimum milk protein in milk solids-not-fat[a]	34% m/m

(a) The milk solids-not-fat content includes water of crystallization of the lactose.

4 FOOD ADDITIVES

Only those food additives listed below may be used and only within limits specified.

INS no.	Name of additive	Maximum level
Stabilizers		
331(i)	Sodium dihydrogen citrate	Limited by GMP
331(iii)	Trisodium citrate	
332(i)	Potassium dihydrogen citrate	
332(ii)	Tripotassium citrate	
508	Potassium chloride	
509	Calcium chloride	
Acidity regulators		
339(i)	Sodium dihydrogen phosphate	4 400 mg/kg, singly or in combination as phosphorous
339(ii)	Disodium hydrogen phosphate	
339(iii)	Trisodium phosphate	
340(i)	Potassium phosphate	
340(ii)	Dipotassium hydrogen phosphate	
340(iii)	Tripotassium phosphate	
341(i)	Calcium dihydrogen phosphate	
341(ii)	Calcium hydrogen phosphate	
450(i)	Disodium diphosphate	
450(ii)	Trisodium diphosphate	
450(iii)	Tetrasodium diphosphate	
450(v)	Tetrapotassium diphosphate	
450(vi)	Dicalcium diphosphate	
450(vii)	Calcium dihydrogen diphosphate	
451(i)	Pentasodium triphosphate	
451(ii)	Pentapotassium triphosphate	
452(i)	Sodium polyphosphate	
452(ii)	Potassium polyphosphate	

STANDARD FOR A BLEND OF SKIMMED MILK AND VEGETABLE FAT IN POWDERED FORM

INS no.	Name of additive	Maximum level
452(iii)	Sodium calcium polyphosphate	4 400 mg/kg, singly or in combination as phosphorous
452(iv)	Calcium polyphosphates	
452(v)	Ammonium polyphosphates	
500(i)	Sodium carbonate	Limited by GMP
500(ii)	Sodium hydrogen carbonate	
500(iii)	Sodium sesquicarbonate	
501(i)	Potassium carbonates	
501(ii)	Potassium hydrogen carbonate	
Emulsifiers		
322	Lecithins	Limited by GMP
471	Mono- and diglycerides of fatty acids	
Anticaking agents		
170(i)	Calcium carbonate	Limited by GMP
504(i)	Magnesium carbonate	
530	Magnesium oxide	
551	Silicon dioxide , amorphous	
552	Calcium silicate	
553(i)	Magnesium silicate, synthetic	
553(iii)	Talc	
554	Sodium aluminium silicate	570 mg/kg, expressed as aluminium
341(iii)	Tricalcium phosphate	4 400 mg/kg, singly or in combination as phosphorous
343(iii)	Trimagnesium phosphate	
Antioxidants		
300	Ascorbic acid , *L*-	500 mg/kg as ascorbic acid
301	Sodium ascorbate	
304	Ascorbyl palmitate	80 mg/kg, singly or in combination as ascorbyl stearate
305	Ascorbyl stearate	
319	Tertiary buthylydroquinone	100 mg/kg singly or in combination. Expressed on fat or oil basis
320	Buthylated hydroxyanisole	
321	Buthylated hydroxytoluene	

5 **CONTAMINANTS**

The products covered by this Standard shall comply with the Maximum Levels for contaminants that are specified for the product in the *General Standard for Contaminants and Toxins in Food and Feed* (CXS 193-1995).

The milk used in the manufacture of the products covered by this Standard shall comply with the Maximum Levels for contaminants and toxins specified for milk by the *General Standard for Contaminants and Toxins in Food and Feed* (CXS 193-1995) and with the maximum residue limits for veterinary drug residues and pesticides established for milk by the CAC.

The vegetable oils/fat used in the manufacture of the products covered by this Standard shall comply with the Maximum Levels for contaminants and toxins specified for the oils/fats by the *General Standard for Contaminants and Toxins in Food and Feed* (CXS 193-1995) and with the maximum residue limits for pesticides established for the oils/fats by the CAC.

6   HYGIENE

It is recommended that the products covered by the provisions of this standard be prepared and handled *in* accordance with the appropriate sections of the *General Principles of Food Hygiene* (CXC 1-1969), the *Code of Hygienic Practice for Milk and Milk Products* (CXC 57-2004) and other relevant Codex texts such as Codes of Hygienic Practice and Codes of Practice. The products should comply with any microbiological criteria established in accordance with the *Principles and Guidelines for the Establishment and Application of Microbiological Criteria Related to Foods* (CXG 21-1997).

7   LABELLING

In addition to the provisions of the *General Standard for the Labelling of Prepackaged Foods* (CXS 1-1985) the following specific provisions apply:

7.1   Name of the food

The name of the food shall be:

- Blend of Skimmed Milk and Vegetable Fat in Powdered Form; or
- Reduced Fat Blend of Skimmed Milk and Vegetable Fat in Powdered Form.

Other names may be used if allowed by national legislation in the country of retail sale.

7.2   Declaration of total fat content

The total fat content shall be declared in a manner found acceptable in the country of sale to the final consumer, either (i) as a percentage by mass or volume, or (ii) in grams per serving as quantified in the label provided that the number of servings is stated.

A statement shall appear on the label as to the presence of edible vegetable fat and/or edible vegetable oil. When required by the country of retail sale, the common name of the vegetable from which the fat or oil is derived shall be included in the name of the food or as a separate statement.

7.3   Declaration of milk protein

The milk protein content shall be declared in a manner acceptable in the country of sale to the final consumer, either (i) as a percentage by mass or volume, or (ii) in grams per serving as quantified in the label provided that the number of servings is stated.

7.4   List of ingredients

Notwithstanding the provision of Section 4.2.1 of the *General Standard for the Labelling*

STANDARD FOR A BLEND OF SKIMMED MILK AND VEGETABLE FAT IN POWDERED FORM

of Prepackaged Foods (CXS 1-1985) milk products used only for protein adjustment need not be declared.

7.5 Advisory statement

A statement shall appear on the label to indicate that the product should not be used as a substitute for infant formula. For example, "NOT SUITABLE FOR INFANTS".

8 METHODS OF SAMPLING AND ANALYSIS

For checking the compliance with this standard, the methods of analysis and sampling contained in the *Recommended Methods of Analysis and Sampling* (CXS 234-1999) relevant to the provisions in this standard, shall be used.

脱脂甜炼乳与植物脂肪混合物

STANDARD FOR A BLEND OF SWEETENED CONDENSED SKIMMED MILK AND VEGETABLE FAT

CXS 252-2006

2006年通过。2010年、2018年修订。

1 范围

本标准适用于符合本标准第2条所述，即食或用于进一步加工的脱脂甜炼乳与植物脂肪混合物。

2 说明

脱脂甜炼乳与植物脂肪混合物是指重组乳成分和饮用水，或进行部分脱水并添加糖、食用植物油、食用植物脂肪及其混合物，满足本标准第3条的成分要求。

3 基本成分和质量要求

3.1 基本原料

脱脂乳和脱脂乳粉[1]，其他非脂乳固体和可食植物脂肪/油[1]。

允许用下列乳制品调整蛋白质含量：

- 乳超滤滞留物：乳超滤滞留物是指乳、部分脱脂乳或脱脂乳，通过超滤法浓缩乳蛋白后形成的产品；
- 乳渗透物：乳渗透物是指通过超滤法，将乳蛋白和乳脂从乳、部分脱脂乳或脱脂乳中分离除去而形成的产品；
- 乳糖[1]：也用于接种的目的。

3.2 其他配料

- 饮用水；
- 糖；
- 氯化钠和/或氯化钾作为替代盐。

在此产品中，糖类一般用蔗糖，但也可按照良好生产规范，将蔗糖与其他糖类混合使用。

---

[1] 规格参见相关国际食品法典标准。

3.3 允许营养素

按照《食品中必需营养素添加通则》（CXG 9-1987）规定，由各个国家立法，按需酌情制定维生素A、维生素D和其他营养素的最高和最低含量，包括禁止使用某些特定营养素。

3.4 成分

脱脂甜炼乳与植物脂肪混合物

总脂肪含量	8% m/m
非脂乳固体最低含量[a]	20% m/m
乳的非脂乳固体中乳蛋白最低含量[a]	34% m/m

低脂、脱脂甜炼乳与植物脂肪混合物

总脂肪最低含量	>1%，<8% m/m
非脂乳固体最低含量[a]	20% m/m
乳的非脂固体中乳蛋白最低含量[a]	34% m/m

（a）乳的非脂固体包括乳糖的结晶水。

4 食品添加剂

只允许在规定范围内使用下表列出的食品添加剂。

INS编号	添加剂名称	最大限量
乳化剂		
322	卵磷脂	根据GMP限量使用
稳定剂		
331（i）	柠檬酸二氢钠	根据GMP限量使用
331（iii）	柠檬酸钠	
332（i）	柠檬酸二氢钾	
332（ii）	柠檬酸钾	
333	柠檬酸钙	
508	氯化钾	
509	氯化钙	
酸度调节剂		
170（i）	碳酸钙	根据GMP限量使用
339（i）	磷酸二氢钠	4 400 mg/kg，单用或混用，以磷计
339（ii）	磷酸氢二钠	
339（iii）	磷酸三钠	
340（i）	磷酸二氢钾	
340（ii）	磷酸氢二钾	

INS编号	添加剂名称	最大限量
340（iii）	磷酸三钾	4 400 mg/kg，单用或混用，以磷计
341（i）	磷酸二氢钙	
341（ii）	磷酸氢二钙	
341（iii）	磷酸三钙	
450（i）	磷酸二钠	
450（ii）	焦磷酸三钠	
450（iii）	焦磷酸四钠	
450（v）	二磷酸四钾	
450（vi）	二磷酸二钙	
450（vii）	二磷酸二氢钙	
451（i）	三磷酸五钠	
451（ii）	三磷酸五钾	
452（i）	六偏磷酸钠	
452（ii）	聚磷酸钾	
452（iii）	聚磷酸钠钙	
452（iv）	聚磷酸钙	
452（v）	聚磷酸铵	根据GMP限量使用
500（i）	碳酸钠	
500（ii）	碳酸氢钠	
500（iii）	倍半碳酸钠	
501（i）	碳酸钾	
501（ii）	碳酸氢钾	
增稠剂		
407	卡拉胶	根据GMP限量使用
407a	加工琼芝属海藻胶（PES）	

5　污染物

本标准所涉及的产品应符合《食品及饲料中污染物和毒素通用标准》（CXS 193-1995）规定的污染物最大限量。

本标准所涉及的产品在加工中使用的牛乳应符合《食品及饲料中污染物和毒素通用标准》（CXS 193-1995）规定的乳中污染物和毒素最大限量以及国际食品法典委员会设定的农药和兽药最大残留限量。

本标准所涉及的产品在生产中所用植物油/脂肪应遵循《食品及饲料中污染物和毒素通用标准》（CXS 193-1995）中规定的油/脂肪中污染物和毒素最大限量以及国际

食品法典委员会设定的油/脂肪中农药最大残留限量。

6 卫生要求

建议本标准所涉及的产品应遵循《食品卫生总则》（CXC 1-1969）、《乳及乳制品卫生操作规范》（CXC 57-2004）以及《卫生操作规范》和《生产操作规范》等其他相关法典文本。产品应符合《食品微生物标准制定与实施原则和准则》（CXG 21-1997）规定的所有微生物标准。

7 标识

除符合《预包装食品标识通用标准》（CXS 1-1985）外，还应符合下列具体规定。

7.1 产品名称

产品名称应为：
- 脱脂甜炼乳与植物脂肪混合物；
- 低脂脱脂甜炼乳与植物脂肪混合物。

如果产品销售国立法允许，也可用其他名称。

7.2 乳脂含量说明

总脂肪含量应以销售国可接受的方式向最终消费者声明：（i）质量百分比或体积百分比形式；（ii）如果产品标识上标明了份数，也可用每份中乳脂重量（g）表示。标识上需说明是否含有可食用植物脂肪或可食用植物油。当销售国有要求时，脂肪或油的植物来源应包含在食品名称中或另附说明。

7.3 乳蛋白含量说明

乳蛋白含量应以销售国可接受的方式向最终消费者声明：（i）质量百分比或体积百分比形式；（ii）如果产品标识上标明了份数，也可用每份中乳蛋白重量（g）表示。

7.4 配料表

尽管《预包装食品标识通用标准》（CXS 1-1985）第4.2.1条有规定，但仅用于调节乳蛋白含量的乳制品不需要声明。

7.5 建议性说明

标识上要有指示该产品不能代替婴儿配方食品的说明。例如"婴儿不适用"。

8 抽样和分析方法

为检查产品是否符合本标准，应采用《分析和抽样推荐性方法》（CXS 234-1999）中涉及本标准规定的分析和抽样方法。

STANDARD FOR A BLEND OF SWEETENED CONDENSED SKIMMED MILK AND VEGETABLE FAT

CXS 252-2006

Adopted in 2006. Amended in 2010, 2018.

1 SCOPE

This Standard applies to a blend of sweetened condensed skimmed milk and vegetable fat, intended for direct consumption, or further processing, in conformity with the description in Section 2 of this Standard.

2 DESCRIPTION

A blend of sweetened condensed skimmed milk and vegetable fat is a product prepared by recombining milk constituents and potable water, or by the partial removal of water, with the addition of sugar and with the addition of edible vegetable oil, edible vegetable fat or a mixture thereof to meet the compositional requirements in Section 3 of this Standard.

3 ESSENTIAL COMPOSITION AND QUALITY FACTORS

3.1 Raw materials

Skimmed milk and skimmed milk powders[1], other non-fat milk solids, and edible vegetable fats/oils[1].

The following milk products are allowed for protein adjustment purposes:

— Milk retentate: Milk retentate is the product obtained by concentrating milk protein by ultra-filtration of milk, partly skimmed milk, or skimmed milk;

— Milk permeate: Milk permeate is the product obtained by removing milk protein and milk fat from milk, partly skimmed milk, or skimmed milk by ultra-filtration; and

— Lactose[1]: Also for seeding purposes.

3.2 Permitted ingredients

— Potable water;
— Sugar;
— Sodium chloride and/or potassium chloride as salt substitute.

In this product, sugar is generally considered to be sucrose, but a combination of sucrose with other sugars, consistent with Good Manufacturing Practice, may be used.

3.3 Permitted nutrients

Where allowed in accordance with the *General Principles for the Addition of Essential Nutrients to Food* (CXG 9-1987), maximum and minimum levels for Vitamins A, D and

---

1 For specification, see relevant Codex Standard.

STANDARD FOR A BLEND OF SWEETENED CONDENSED SKIMMED MILK AND VEGETABLE FAT

other nutrients, where appropriate, should be laid down by national legislation in accordance with the needs of individual country including, where appropriate, the prohibition of the use of particular nutrients.

3.4 Composition

Blend of sweetened condensed skimmed milk and vegetable fat

Minimum total fat	8% m/m
Minimum milk solids-not-fat[a]	20% m/m
Minimum milk protein in milk solids-not-fat[a]	34% m/m

Reduced fat blend of sweetened condensed skimmed milk and vegetable fat

Total fat	More than 1% and less than 8% m/m
Minimum milk solids-not-fat[a]	20% m/m
Minimum milk protein in milk solids-not-fat[a]	34% m/m

(a) The milk solids-not-fat content includes water of crystallization of the lactose.

For a blend of sweetened condensed skimmed milk and vegetable fat the amount of sugar is restricted by Good Manufacturing Practice to a minimum value which safeguards the keeping quality of the product and a maximum value above which crystallization of sugar, may occur.

4 FOOD ADDITIVES

Only those food additives listed below may be used and only within the limits specified.

INS no.	Name of additive	Maximum level
Emulsifiers		
322	Lecithins	Limited by GMP
Stabilizers		
331(i)	Sodium dihydrogen citrate	Limited by GMP
331(iii)	Trisodium citrate	Limited by GMP
332(i)	Potassium dihydrogen citrate	Limited by GMP
332(ii)	Tripotassium citrate	Limited by GMP
333	Calcium citrate	Limited by GMP
508	Potassium chloride	Limited by GMP
509	Calcium chloride	Limited by GMP
Acidity regulators		
170(i)	Calcium carbonate	Limited by GMP
339(i)	Sodium dihydrogen phosphate	4 400 mg/kg, singly or in combination as phosphorous
339(ii)	Disodium hydrogen phosphate	

INS no.	Name of additive	Maximum level
339(iii)	Trisodium phosphate	4 400 mg/kg, singly or in combination as phosphorous
340(i)	Potassium dihydrogen phosphate	
340(ii)	Dipotassium hydrogen phosphate	
340(iii)	Tripotassium phosphate	
341(i)	Calcium dihydrogen phosphate	
341(ii)	Calcium hydrogen phosphate	
341(iii)	Tricalcium phosphate	
450(i)	Disodium diphosphate	
450(ii)	Trisodium diphosphate	
450(iii)	Tetrasodium diphosphate	
450(v)	Tetrapotassium diphosphate	
450(vi)	Dicalcium diphosphate	
450(vii)	Calcium dihydrogen diphosphate	
451(i)	Pentasodium triphosphate	
451(ii)	Pentapotassium triphosphate	
452(i)	Sodium polyphosphate	
452(ii)	Potassium polyphosphate	
452(iii)	Sodium calcium polyphosphate	
452(iv)	Calcium polyphosphates	
452(v)	Ammonium polyphosphates	
500(i)	Sodium carbonate	Limited by GMP
500(ii)	Sodium hydrogen carbonate	
500(iii)	Sodium sesquicarbonate	
501(i)	Potassium carbonates	
501(ii)	Potassium hydrogen carbonate	
Thickeners		
407	Carrageenan	Limited by GMP
407a	Processed euchema seaweed (PES)	

STANDARD FOR A BLEND OF SWEETENED CONDENSED SKIMMED MILK AND VEGETABLE FAT

5 CONTAMINANTS

The products covered by this Standard shall comply with the maximum levels for contaminants that are specified for the product in the *General Standard for Contaminants and Toxins in Food and Feed* (CXS 193-1995).

The milk used in the manufacture of the products covered by this Standard shall comply with the maximum levels for contaminants and toxins specified for milk by the *General Standard for Contaminants and Toxins in Food and Feed* (CXS 193-1995) and with the maximum residue limits for veterinary drug residues and pesticides established for milk by the CAC.

The vegetable oils/fat used in the manufacture of the products covered by this Standard shall comply with the Maximum Levels for contaminants and toxins specified for the oils/fats by the *General Standard for Contaminants and Toxins in Food and Feed* (CXS 193-1995) and with the maximum residue limits for pesticides established for the oils/fats by the CAC.

6 HYGIENE

It is recommended that the products covered by the provisions of this Standard be prepared and handled in accordance with the appropriate sections of the *General Principles of Food Hygiene* (CXC 1-1969), the *Code of Hygienic Practice for Milk and Milk Products* (CXC 57-2004) and other relevant Codex texts such as Codes of Hygienic Practice and Codes of Practice. The products should comply with any microbiological criteria established in accordance with the *Principles and Guidelines for the Establishment and Application of Microbiological Criteria Related to Foods* (CXG 21-1997).

7 LABELLING

In addition to the provisions of the *General Standard for the Labelling of Prepackaged Foods* (CXS 1-1985) the following specific provisions apply:

7.1 Name of the food

The name of the food shall be:

- Blend of Sweetened Condensed Skimmed Milk and Vegetable Fat; or
- Reduced Fat Blend of Sweetened Condensed Skimmed Milk and Vegetable Fat.

Other names may be used if allowed by national legislation in the country of retail sale.

7.2 Declaration of total fat content

The total fat content shall be declared in a manner found acceptable in the country of sale to the final consumer, either (i) as a percentage by mass or volume, or (ii) in grams per serving as quantified in the label provided that the number of servings is stated.

A statement shall appear on the label as to the presence of edible vegetable fat and/or edible vegetable oil. When required by the country of retail sale, the common name of the vegetable from which the fat or oil is derived shall be included in the name of the food or as a separate statement.

7.3 Declaration of milk protein

The milk protein content shall be declared in a manner acceptable in the country of sale to the final consumer, either (i) as a percentage by mass or volume, or (ii) in grams per serving as quantified in the label provided that the number of servings is stated.

7.4 List of ingredients

Notwithstanding the provision of Section 4.2.1 of the *General Standard for the Labelling of Prepackaged Foods* (CXS 1-1985) milk products used only for protein adjustment need not be declared.

7.5 Advisory statement

A statement shall appear on the label to indicate that the product should not be used as a substitute for infant formula. For example, "NOT SUITABLE FOR INFANTS".

8 METHODS OF SAMPLING AND ANALYSIS

For checking the compliance with this Standard, the methods of analysis and sampling contained in the *Recommended Methods of Analysis and Sampling* (CXS 234-1999) relevant to the provisions in this Standard, shall be used.

乳脂涂抹物

STANDARD FOR DAIRY FAT SPREADS

CXS 253-2006

2006年通过。2008年、2010年、2018年修订。

1 范围

本标准适用于符合本标准第2条所述即食或用于进一步加工的乳脂涂抹物。

2 说明

乳脂涂抹物是指一种相对富含脂肪的乳制品，呈可涂抹乳剂状，主要为乳包水乳化形式，在20 ℃时仍然保持固态。

3 基本成分和质量要求

3.1 基本原料

– 乳和/或乳制品。

包括乳脂在内的原材料，可在使用前进行一切适当加工（例如，物理改性处理，包括分馏）。

3.2 其他配料

可以添加下列物质：

– 调味剂；
– 安全、适宜的加工助剂；
– 按照《食品中必需营养素添加通则》（CXG 9-1987）相关规定，根据个别国家需求，由国家立法按需制定维生素A、维生素D和其他营养素最高和最低限量，包括禁止使用某些特定营养素；
– 作为代盐制品的氯化钠和氯化钾；
– 糖（一切碳水化合物甜味剂）；
– 菊糖和麦芽糊精（根据GMP限量使用）；
– 无害乳酸发酵剂和/或产香菌；
– 水；
– 明胶和淀粉（根据GMP限量使用）。这两种物质可以替代增稠剂发挥相同功能，但需斟酌第4条所列增稠剂的一切用法，以GMP为准，添加剂量仅以发挥必要的功能为限。

3.3 成分

乳脂含量范围需在10% ~ 80%（m/m），并且至少占干物质含量的2/3。

乳脂涂抹物的成分调整以《乳品术语》（CXS 206-1999）第4.3.3条规定为准。

4 食品添加剂

只允许在规定范围内使用下表列出的食品添加剂。

功能类别	在乳脂中合理使用	
	<70%乳脂(a)	≥70%乳脂
酸度调节剂	×	×
抗结块剂	—	—
消泡剂	×	×
抗氧化剂	×	×
漂白剂	—	—
疏松剂	—	—
碳酸化剂	—	—
着色剂	×	×
护色剂	—	—
乳化剂	×	—
固化剂	—	—
增味剂	×	—
发泡剂	—	—
胶凝剂	—	—
保湿剂	—	—
防腐剂	×	×
推进剂	×	×
膨松剂	—	—
螯合剂	—	—
稳定剂	×	—
增稠剂	×	—

（a）乳化剂、稳定剂、增稠剂和增味剂的使用，要根据GMP限量添加，考虑到随着脂肪含量增加，达到工艺上的功能所需添加的量可以减少。在脂肪含量约70%时几乎无须使用。

INS编号	添加剂名称	最大限量
着色剂		
100（i）	姜黄素	5 mg/kg
160a（i）	β-胡萝卜素（人工合成）	35 mg/kg，单用或混用
160a（ii）	β-胡萝卜素（天然萃取物）	
160e	β-apo-8′-胡萝卜素醛	
160f	β-apo-8′-胡萝卜酸素乙酯	
160b（i）	胭脂树提取物-胭脂素	20 mg/kg
乳化剂		
432	聚氧乙烯（20）山梨糖醇酐单月桂酸酯	10 000 mg/kg，单用或混用（仅用于烘烤的乳脂涂抹物）
433	聚氧乙烯（20）山梨糖醇酐单油酸酯	
434	聚氧乙烯（20）山梨糖醇酐单棕榈酸酯	
435	聚氧乙烯（20）山梨糖醇酐单硬脂酸酯	
436	聚氧乙烯（20）山梨糖醇酐三硬脂酸酯	
471	单、双甘油脂肪酸酯	根据GMP限量使用
472a	乙酰化单、双甘油脂肪酸酯	
472b	乳酸脂肪酸甘油酯	
472c	柠檬酸脂肪酸甘油酯	
472e	二乙酰酒石酸酯和脂肪酸酯	10 000 mg/kg
473	蔗糖脂肪酸酯	10 000 mg/kg，仅用于烘烤的乳脂涂抹物
474	蔗糖甘油酯	
475	聚甘油脂肪酸酯	5 000 mg/kg
476	聚甘油蓖麻醇酸酯	4 000 mg/kg
481（i）	硬脂酰乳酸钠	10 000 mg/kg，单用或混用
482（i）	硬脂酰乳酸钙	
491	山梨醇酐单硬脂酸酯	10 000 mg/kg，单用或混用
492	山梨醇酐三硬脂酸酯	
493	山梨醇酐单月桂酸酯	
494	山梨醇酐单油酸酯	
495	山梨醇酐单棕榈酸酯	

INS编号	添加剂名称	最大限量
防腐剂		
200	山梨酸	脂肪含量<59%：2 000 mg/kg，单用或混合使用（以山梨酸计） 脂肪含量≥59%：1 000 mg/kg，单用或混合使用（以山梨酸计）
202	山梨酸钾	
203	山梨酸钙	
稳定剂/增稠剂		
340（i）	磷酸二氢钾	880 mg/kg，单用或混用，以磷计
340（ii）	磷酸氢二钾	
340（iii）	磷酸三钾	
341（i）	磷酸二氢钙	
341（ii）	磷酸氢二钙	
341（iii）	磷酸三钙	
450（i）	磷酸二钠	
400	海藻酸	根据GMP限量使用
401	海藻酸钠	
402	海藻酸钾	
403	海藻酸铵	
404	海藻酸钙	
406	琼脂	
405	海藻酸丙二醇酯	3 000 mg/kg
407	卡拉胶	根据GMP限量使用
407a	加工琼芝属海藻胶（PES）	
410	槐豆胶（又名刺槐豆胶）	
412	瓜尔胶	
413	黄蓍胶	
414	阿拉伯胶	
415	黄原胶	
418	结冷胶	
422	甘油	
440	果胶	
460（i）	微晶纤维素	
460（ii）	纤维素粉	

INS编号	添加剂名称	最大限量
461	甲基纤维素	根据GMP限量使用
463	羟丙基纤维素	
464	羟丙基甲基纤维素	
465	甲基乙基纤维素	
466	羧甲基纤维素钠（纤维素胶）	
500（i）	碳酸钠	
500（ii）	碳酸氢钠	
500（iii）	倍半碳酸钠	
1400	糊精，焙炒淀粉	
1401	酸处理淀粉	
1402	碱处理淀粉	
1403	漂白淀粉	
1404	氧化淀粉	
1405	酶处理淀粉	
1410	单淀粉磷酸酯	
1412	磷酸酯双淀粉	
1413	磷酸化二淀粉磷酸酯	
1414	乙酰化二淀粉磷酸酯	
1420	醋酸酯淀粉	
1422	乙酰化双淀粉己二酸酯	
1440	羟丙基淀粉	
1442	羟丙基二淀粉磷酸酯	
酸度调节剂		
325	乳酸钠	根据GMP限量使用
326	乳酸钾	
327	乳酸钙	
329	DL-乳酸镁	
331（i）	柠檬酸二氢钠	
331（ii）	柠檬酸二钠	
334	$L(+)$-酒石酸	5 000 mg/kg，单用或混用，以酒石酸计
335（ii）	$L(+)$-酒石酸钠	

INS编号	添加剂名称	最大限量
337	L(+)-酒石酸钾钠	
339（i）	磷酸二氢钠	880 mg/kg，单用或混用，以磷计
339（ii）	磷酸氢二钠	
339（iii）	磷酸三钠	
338	磷酸	
524	氢氧化钠	根据GMP限量使用
526	氢氧化钙	
抗氧化剂		
304	抗坏血酸棕榈酸酯	500 mg/kg，以抗坏血酸硬脂酸酯计
305	抗坏血酸硬脂酸酯	
307	生育酚	500 mg/kg
310	没食子酸丙酯	200 mg/kg，单用或混合使用：丁基羟基茴香醚（BHA）（INS 320），丁基羟基甲苯（BHT）（INS 321）和没食子酸丙酯（INS 310），以脂肪或油脂为基础的最大混合用量200 mg/kg计。仅可用于烹调用的乳脂涂抹物
320	丁基羟基茴香醚	
321	丁基羟基甲苯	75 mg/kg，单用或混合使用：丁基羟基茴香醚（BHA）（INS 320），丁基羟基甲苯（BHT）（INS 321）和没食子酸丙酯（INS 310），以脂肪或油脂为基础的最大混合用量200 mg/kg计。仅可用于烹调用的乳脂涂抹物
消泡剂		
900a	聚二甲基硅氧烷	10 mg/kg，仅用于煎炸用途的乳脂涂抹物
增味剂		
627	5′-鸟苷酸二钠	根据GMP限量使用
628	5′-鸟苷酸二钾	

5　污染物

本标准所涉及的产品应符合《食品及饲料中污染物和毒素通用标准》（CXS 193-1995）规定的污染物最大限量。

本标准所涉及的产品在加工中使用的牛乳应符合《食品及饲料中污染物和毒素通用标准》（CXS 193-1995）规定的乳中污染物和毒素最大限量以及国际食品法典委员会设定的农药和兽药最大残留限量。

6 卫生要求

建议本标准所涉及的产品应遵循《食品卫生总则》（CXC 1-1969）、《乳及乳制品卫生操作规范》（CXC 57-2004）以及《卫生操作规范》和《生产操作规范》等其他相关法典文本。产品应符合《食品微生物标准制定与实施原则和准则》（CXG 21-1997）规定的所有微生物标准。

7 标识

除符合《预包装食品标识通用标准》（CXS 1-1985）外，还应符合下列具体规定。

7.1 产品名称

7.1.1 产品名称为"乳脂涂抹物"。如果产品销售国立法允许，也可用其他名称。

7.1.2 根据《营养和保健声明使用准则》（CXG 23-1997），低脂乳脂涂抹物可以标注为"低脂"。

7.1.3 名称和一切形容词，在翻译成其他语言时应避免产生歧义，不必逐字翻译，并应符合产品销售国语言习惯。

7.1.4 乳脂涂抹物的标识，还要根据国家立法规定，说明其是否加盐。

7.1.5 加糖乳脂涂抹物应标注加糖。

7.2 乳脂含量说明

乳脂含量应以销售国可接受的方式向最终消费者声明：（i）质量百分比或体积百分比形式；（ii）如果产品标识上标明了份数，也可用每份中乳脂重量（g）表示。

7.3 非零售包装标识

在包装容器上除要标明产品名称、批次、生产厂家或包装商的名称和地址外，在包装容器上或附随说明书中，也应按本标准第7条和《预包装食品标识通用标准》（CXS 1-1985）第4.1~4.8条所要求的信息以及储存说明（在必要的地方）加以陈述。如果批次、生产厂家或包装商的名称和地址可以在附随的文件标明，则可以用一个识别标识来代替。

8 抽样和分析方法

为检查产品是否符合本标准，应采用《分析和抽样推荐性方法》（CXS 234-1999）中涉及本标准规定的分析和抽样方法。

STANDARD FOR DAIRY FAT SPREADS

CXS 253-2006

Adopted in 2006. Amended in 2008, 2010, 2018.

1 SCOPE

This Standard applies to dairy fat spreads intended for use as spreads for direct consumption, or for further processing, in conformity with Section 2 of this Standard.

2 DESCRIPTION

Dairy fat spreads are milk products relatively rich in fat in the form of a spreadable emulsion principally of the type of water-in-milk fat that remains in solid phase at a temperature of 20 °C.

3 ESSENTIAL COMPOSITION AND QUALITY FACTORS

3.1 Raw materials

— Milk and/or products obtained from milk.

Raw materials, including milk fat, may have been subjected to any appropriate processing (e.g. physical modifications including fractionation) prior to its use.

3.2 Permitted ingredients

The following substances may be added:

- Flavours and flavourings;
- Safe and suitable processing aids;
- Where allowed in accordance with the *General Principles for the Addition of Essential Nutrients to Food* (CXG 9-1987), maximum and minimum levels for vitamins A, D and other nutrients, where appropriate, should be laid down by national legislation in accordance with the needs of individual countries including, where appropriate, the prohibition of the use of particular nutrients;
- Sodium chloride and potassium chloride as a salt substitute;
- Sugars (any carbohydrate sweetening matter);
- Inulin and malto-dextrins (limited by GMP);
- Starter cultures of harmless lactic acid and/or flavour producing bacteria;
- Water;
- Gelatine and Starches (limited by GMP). These substances can be used in the same function as thickeners, provided they are added only in amounts functionally necessary as governed by GMP taking into account any use of the thickeners listed in Section 4.

3.3 Composition

The milk fat content shall be no less than 10% and less than 80% (m/m) and shall represent at least 2/3 of the dry matter.

Compositional modifications of Dairy Fat Spreads are restricted by the requirements of Section 4.3.3 of the *General Standard for the Use of Dairy Terms* (CXS 206-1999).

4 FOOD ADDITIVES

Only those additive functional classes indicated as technologically justified in the table below may be used for the product categories specified. Within each additive class, and where permitted according to the table, only those food additives listed below the table may be used and only within the functions and limits specified.

Additive functional class	Justified use in dairy fat spreads:	
	< 70% milk fat content[a]	≥ 70% milk fat content
Acidity regulators	×	×
Anticaking agents	–	–
Antifoaming agents	×	×
Antioxidants	×	×
Bleaching agents	–	–
Bulking agents	–	–
Carbonating agents	–	–
Colours	×	×
Colour retention agents	–	–
Emulsifiers	×	–
Firming agents	–	–
Flavour enhancers	×	–
Foaming agents	–	–
Gelling agents	–	–
Humectants	–	–
Preservatives	×	×
Propellants	×	×
Raising agents	–	–
Sequestrants	–	–
Stabilizers	×	–
Thickeners	×	–

(a) The application of GMP in the use of emulsifiers, stabilizers, thickeners and flavour enhancers includes consideration of the fact that the amount required to obtain the technological function in the product decreases with increasing fat content, fading out at fat content about 70%.

INS no.	Name of additive	Maximum level
Colours		
100(i)	Curcumin	5 mg/kg
160a(i)	Carotene, *beta-*,synthetic	35 mg/kg, singly or in combination
160a(ii)	Carotene, *beta-*,*Blakeslea trispora*	
160e	Carotenal,*beta-apo*-8′-	
160f	Carotenoic acid, ethyl ester *beta-apo*-8′-	
160b(i)	Annatto extracts- bixin based	20 mg/kg
Emulsifiers		
432	Polyoxyethylene (20) sorbitan monolaurate	10 000 mg/kg, singly or in combination (Dairy fat spreads for baking purposes only)
433	Polyoxyethylene (20) sorbitan monooleate	
434	Polyoxyethylene (20) sorbitan monopalmitate	
435	Polyoxyethylene (20) sorbitan monostearate	
436	Polyoxyethylene (20) sorbitan tristearate	
471	Mono and diglycerides of fatty acids	Limited by GMP
472a	Acetic and fatty acid esters of glycerol	
472b	Lactic and fatty acid esters of glycerol	
472c	Citric and fatty acid esters of glycerol	
472e	Diacetyltartaric and fatty acid esters of glycerol	10 000 mg/kg
473	Sucrose esters of fatty acids	10 000 mg/kg, dairy fat spreads for baking purposes only
474	Sucroglycerides	10 000 mg/kg, dairy fat spreads for baking purposes only
475	Polyglycerol esters of fatty acids	5 000 mg/kg
476	Polyglycerol esters of interesterified ricinoleic acid	4 000 mg/kg
481(i)	Sodium stearoyl lactylate	10 000 mg/kg, singly or in combination
482(i)	Calcium stearoy lactylate	
491	Sorbitan monostearate	10 000 mg/kg, singly or in combination
492	Sorbitan tristearate	
493	Sorbitan monolaurate	

INS no.	Name of additive	Maximum level
494	Sorbitan monooleate	10 000 mg/kg, singly or in combination
495	Sorbitan monopalmitate	
Preservatives		
200	Sorbic acid	2 000 mg/kg, singly or in combination (as sorbic acid) for fat contents < 59% and 1 000 mg/kg singly or in combination (as sorbic acid) for fat contents $\geqslant$ 59%
202	Potassium sorbate	
203	Calcium sorbate	
Stabilizers/thickeners		
340(i)	Potassium dihydrogen phosphate	880 mg/kg, singly or in combination, as phosphorous
340(ii)	Dipotassium hydrogen phosphate	
340(iii)	Tripotassium phosphate	
341(i)	Calcium dihydrogen sphosphate	
341(ii)	Calcium hydrogen phosphate	
341(iii)	Tricalcium phosphate	
450(i)	Disodium diposphate	
400	Alginic acid	Limited by GMP
401	Sodium alginate	
402	Potassium alginate	
403	Ammonium alginate	
404	Calcium alginate	
406	Agar	
405	Propylene glycol alginate	3 000 mg/kg
407	Carrageenan	Limited by GMP
407a	Processed euchema seaweed (PES)	
410	Carob bean gum	
412	Guar gum	
413	Tragacanth gum	
414	Gum arabic (acacia gum)	
415	Xanthan gum	
418	Gellan gum	

INS no.	Name of additive	Maximum level
422	Glycerol	Limited by GMP
440	Pectins	
460(i)	Microcrystalline cellulose (Cellulose gel)	
460(ii)	Powdered cellulose	
461	Methyl cellulose	
463	Hydroxypropyl cellulose	
464	Hydroxypropyl methyl cellulose	
465	Methyl ethyl cellulose	
466	Sodium carboxymethyl cellulose (Cellulose gum)	
500(i)	Sodium carbonate	
500(ii)	Sodium hydrogen carbonate	
500(iii)	Sodium sesquicarbonate	
1400	Dextrin, roasted starch	
1401	Acid-treated starch	
1402	Alkaline-treated starch	
1403	Bleached starch	
1404	Oxidized starch	
1405	Starches, enzyme treated	
1410	Monostarch phosphate	
1412	Distarch phosphate	
1413	Phosphated distarch phosphate	
1414	Acetylated distarch phosphate	
1420	Starch acetate esterified with acetic anhydride	
1422	Acetylated distarch adipate	
1440	Hydroxypropyl starch	
1442	Hydroxypropyl distarch phosphate	
Acidity regulators		
325	Sodium lactate	Limited by GMP
326	Potassium lactate	

STANDARD FOR DAIRY FAT SPREADS

INS no.	Name of additive	Maximum level
327	Calcium lactate	Limited by GMP
329	Magnesium lactate, *DL*-	
331(i)	Sodium dihydrogen citrate	
331(ii)	Disodium monohydrogen citrate	
334	Tartaric acid, *L*(+)-	5 000 mg/kg, singly or in combination as tartaric acid
335 (ii)	Disodium tartrate	
337	Potassium sodium (*L*+)-tartrate	
339 (i)	Sodium dihydrogen phosphate	880 mg/kg, singly or in combination as phosphorous
339 (ii)	Sodium hydrogen phosphate	
339 (iii)	Trisodium phosphate	
338	Phosphoric acid	
524	Sodium hydroxide	Limited by GMP
526	Calcium hydroxide	
Antioxidants		
304	Ascorbyl palmitate	500 mg/kg, as ascorbyl stearate
305	Ascorbyl stearate	
307	Tocopherols	500 mg/kg
310	Propyl gallate	200 mg/kg, singly or in combination: butylated hydroxyanisole (INS 320), butylated hydroxytoluene (INS 321), and propyl gallate (INS 310) as a combined maximum level of 200 mg/kg on a fat or oil basis. May be used only in dairy fat spreads intended for cooking purposes.
320	Butylated hydroxyanisole	200 mg/kg, singly or in combination: Butylated hydroxyanisole (INS 320), butylated hydroxytoluene (INS 321), and propyl gallate (INS 310) as a combined maximum level of 200 mg/kg on a fat or oil basis. May be used only in dairy fat spreads intended for cooking purposes.

INS no.	Name of additive	Maximum level
321	Butylated hydroxytoluene	75 mg/kg, singly or in combination: butylated hydroxyanisole (INS 320), butylated hydroxytoluene (INS 321), and propyl gallate (INS 310) as a combined maximum level of 200 mg/kg on a fat or oil basis. May be used only in dairy fat spreads intended for cooking purposes.
Anti-foaming agents		
900a	Polydimethylsiloxane	10 mg/kg in dairy fat spreads for frying purposes, only
Flavour enhancers		
627	Disodium 5′-guanylate	Limited by GMP
628	Dipotassium 5′-guanylate	

5 CONTAMINANTS

The products covered by this Standard shall comply with the Maximum Levels for contaminants that are specified for the product in the *General Standard for Contaminants and Toxins in Food and Feed* (CXS 193-1995).

The milk used in the manufacture of the products covered by this Standard shall comply with the Maximum Levels for contaminants and toxins specified for milk by the *General Standard for Contaminants and Toxins in Food and Feed* (CXS 193-1995) and with the maximum residue limits for veterinary drug residues and pesticides established for milk by the CAC.

6 HYGIENE

It is recommended that the products covered by the provisions of this standard be prepared and handled in accordance with the appropriate sections of the *General Principles of Food Hygiene* (CXC 1-1969), the *Code of Hygienic Practice for Milk and Milk Products* (CXC 57-2004) and other relevant Codex texts such as Codes of Hygienic Practice and Codes of Practice. The products should comply with any microbiological criteria established in accordance with the *Principles and Guidelines for the Establishment and Application of Microbiological Criteria Related to Foods* (CXG 21-1997).

7 LABELLING

In addition to the provisions of the *General Standard for the Labelling of Prepackaged Foods* (CXS 1-1985) and the *General Standard for the Use of Dairy Terms* (CXS 206-1999), the following specific provisions apply.

7.1 Name of the food

7.1.1 The name of the food shall be "Dairy Fat Spread". Other names may be used if allowed by national legislation in the country of retail sale.

7.1.2 Dairy fat spreads with reduced fat content may be labelled as "reduced fat" in line with the *Guidelines for Use of Nutrition and Health Claims* (CXG 23-1997).

7.1.3 The designations and any qualifying terms should be translated into other languages in a non-misleading way and not necessarily word for word and should be acceptable in the country of retail sale.

7.1.4 Dairy fat spread may be labelled to indicate whether it is salted or unsalted according to national legislation.

7.1.5 Dairy fat spreads that have been sweetened shall be labelled to indicate that they have been sweetened.

7.2 **Declaration of fat content**

The milk fat content shall be declared in a manner found acceptable in the country of retail sale, either (i) as a percentage by mass, or (ii) in grams per serving as quantified in the label provided that the number of servings is stated.

7.3 **Labelling of non-retail containers**

Information required in Section 7 of this Standard and Sections 4.1 to 4.8 of the *General Standard for the Labelling of Prepackaged Foods* (CXS 1-1985) and, if necessary, storage instructions, shall be given either on the container or in accompanying documents, except that the name of the product, lot identification and the name and address of the manufacturer or packer shall appear on the container. However, lot identification and the name and address of the manufacturer or packer may be replaced by an identification mark, provided that such a mark is clearly identifiable on the accompanying documents.

8 **METHODS OF SAMPLING AND ANALYSIS**

For checking the compliance with this standard, the methods of analysis and sampling contained in the *Recommended Methods of Analysis and Sampling* (CXS 234-1999) relevant to the provisions in this standard, shall be used.

马苏里拉干酪

STANDARD FOR MOZZARELLA

CXS 262-2006

2006年通过。2010年、2013年、2016年、2018年、2019年修订。

1 范围

本标准适用于符合本标准第2条所述即食或用于进一步加工的马苏里拉干酪。

2 说明

马苏里拉干酪是指符合《干酪通则》（CXS 283-1978）和《未成熟干酪（包括新鲜干酪）标准》（CXS 221-2001）的未成熟硬质干酪。干酪平滑且有弹性，有多股长而平行的纤维蛋白结构，无凝乳颗粒。干酪无外皮[1]并可以制成各种形状。

含水量高的马苏里拉干酪是覆盖乳状液体层的软质干酪。这些覆盖层可能形成袋状，容纳乳状液。干酪在包装时，有无乳状液均可。干酪近乎白色。

含水量低的硬质或半硬质马苏里拉干酪无气孔，适合切碎。

马苏里拉干酪由"帕斯塔菲拉塔"加工工艺制作而成。即加热合适pH值的凝乳，揉捏或拉伸直到凝块平滑无结块。将凝乳趁热切割并入模成型，后冷却凝固。也可采用其他使成品干酪具有相同物理、生化和感官特性的加工工艺。

3 基本成分和质量要求

3.1 基本原料

牛乳、水牛乳或它们的混合乳，以及源于这些乳的制品。

3.2 其他配料

- 无害乳酸发酵剂和/或产香菌和其他无害微生物的培养剂；
- 凝乳酶或其他安全、适宜的凝固酶；
- 氯化钠和氯化钾为盐替代品；
- 安全、适宜的加工助剂；
- 醋；
- 饮用水；
- 大米、玉米、马铃薯粉和淀粉：尽管有《干酪通则》（CXS 283-1978），但这些物质同样可作为抗结块剂处理切块、切片和切碎产品表面，只要添加量是

---

[1] 干酪采用不形成外皮的方式保存（"无外皮"干酪）。

良好操作规范所要求达到期待功能的必需量，同时考虑第4条规定的抗结块剂的使用。

3.3 成分

乳成分	最低含量（m/m）	最高含量（m/m）	参考含量（m/m）
干物质乳脂：高水分	20%	/	40%～50%
干物质乳脂：低水分	18%	/	40%～50%
干物质	根据干物质脂肪含量，参见下表		
	干物质脂肪含量（m/m）	相应最低干物质含量（m/m）	
		低水分	高水分
	≥18%，<30%	34%	−
	≥20%，<30%	−	24%
	≥30%，<40%	39%	26%
	≥40%，<45%	42%	29%
	≥45%，<50%	45%	31%
	≥50%，<60%	47%	34%
	≥60%，<85%	53%	38%

乳脂和干物质含量超过规定的最大值和最小值被视为不符合《乳品术语》（CXS 206-1999）第4.3.3条。

4 食品添加剂

只允许在规定范围内使用下表列出的食品添加剂。

添加剂功能类别	合理使用			
	低水分含量		高水分含量	
	干酪量	表面处理	干酪量	表面处理
着色剂	×[a]	−	×[a]	−
漂白剂	−	−	−	−
酸度调节剂	×	−	×	−
稳定剂	×	−	×	−
增稠剂	×	−	×	−

添加剂功能类别	合理使用			
	低水分含量		高水分含量	
	干酪量	表面处理	干酪量	表面处理
乳化剂	–	–	–	–
抗氧化剂	–	–	–	–
防腐剂	×	×	×	×(c)
发泡剂	–	–	–	–
抗结块剂	–	×(b)	–	×(d)

（a）仅为获得第2条所述颜色特征。
（b）仅用于切片、切块、切碎或磨碎的乳酪表面。
（c）仅用于非液体包装的高水分莫萨里拉干酪。
（d）仅用于切碎和/或切块干酪的表面处理。
× 使用该类别添加剂符合技术要求。
– 使用该类别添加剂不符合技术要求。

INS编号	添加剂名称	最大限量
防腐剂		
200	山梨酸	1 000 mg/kg，单用或混用，以山梨酸计
202	山梨酸钾	
203	山梨酸钙	
234	乳酸链球菌素	12.5 mg/kg
235	纳他霉素（匹马菌素）	小于2 mg/dm^2，且5 mm深处不存在
280	丙酸	根据GMP限量使用
281	丙酸钠	
282	丙酸钙	
283	丙酸钾	
酸度调节剂		
170（i）	碳酸钙	根据GMP限量使用
260	冰醋酸	
261（i）	醋酸钾	
261（ii）	双乙酸钾	

INS编号	添加剂名称	最大限量
262（i）	乙酸钠	根据GMP限量使用
263	乙酸钙	
270	乳酸（L-，D-和DL-）	
296	DL-苹果酸	
325	乳酸钠	
326	乳酸钾	
327	乳酸钙	
330	柠檬酸	
338	磷酸	880 mg/kg，以磷计
350（i）	DL-苹果酸氢钠	根据GMP限量使用
350（ii）	苹果酸钠	
352（ii）	D，L-苹果酸钙	
500（i）	碳酸钠	
500（ii）	碳酸氢钠	
500（iii）	倍半碳酸钠	
501（i）	碳酸钾	
501（ii）	碳酸氢钾	
504（i）	碳酸镁	
504（ii）	碳酸氢镁	
507	盐酸	
575	葡萄糖酸-δ-内酯	
577	葡萄糖酸钾	
578	葡萄糖酸钙	
稳定剂		
331（i）	柠檬酸二氢钠	根据GMP限量使用
332（i）	柠檬酸二氢钾	
333	柠檬酸钙	
339（i）	磷酸二氢钠	4 400 mg/kg，单用或混用，以磷计
339（ii）	磷酸氢二钠	
339（iii）	磷酸三钠	
340（i）	磷酸二氢钾	

INS编号	添加剂名称	最大限量
340（ii）	磷酸氢二钾	4 400 mg/kg，单用或混用，以磷计
340（iii）	磷酸三钾	
341（i）	磷酸二氢钙	
341（ii）	磷酸氢二钙	
341（iii）	磷酸三钙	
342（i）	磷酸二氢铵	
342（ii）	磷酸氢二铵	
343（ii）	磷酸氢二镁	
343（iii）	磷酸三镁	
450（i）	磷酸二钠	
450（iii）	焦磷酸四钠	
450（v）	二磷酸四钾	
450（vi）	磷酸二钙	
451（i）	三磷酸五钠	
451（ii）	三磷酸五钾	
452（i）	六偏磷酸钠	
452（ii）	聚磷酸钾	
452（iv）	聚磷酸钙	
452（v）	聚磷酸铵	
406	琼脂	根据GMP限量使用
407	卡拉胶	
407（a）	加工琼芝属海藻胶（PES）	
410	槐豆胶（又名刺槐豆胶）	
412	瓜尔胶	
413	黄蓍胶	
415	黄原胶	
416	刺梧桐胶	
417	刺云实胶	
440	果胶	
466	羧甲基纤维素钠（纤维素胶）	

INS编号	添加剂名称	最大限量
着色剂		
140	叶绿素	根据GMP限量使用
141（i）	叶绿素铜络合物	5 mg/kg，单用或混用
141（ii）	叶绿素铜络合物，钠盐和钾盐	
171	二氧化钛	根据GMP限量使用
抗结块剂		
460（i）	微晶纤维素	根据GMP限量使用
460（ii）	纤维素粉	
551	二氧化硅（无定型）	10 000 mg/kg 单用或混用，以二氧化硅计
552	硅酸钙	
553（i）	硅酸镁（合成）	

* 干酪表面和外皮的定义请参阅《干酪通则》（CXS 283-1978）的附录。

5 污染物

本标准所涉及的产品应符合《食品及饲料中污染物和毒素通用标准》（CXS 193-1995）规定的污染物最大限量。

本标准所涉及的产品在加工中使用的牛乳应符合《食品及饲料中污染物和毒素通用标准》（CXS 193-1995）规定的乳中污染物和毒素最大限量以及国际食品法典委员会设定的农药和兽药最大残留限量。

6 卫生要求

建议本标准所涉及产品应遵循《食品卫生总则》（CXC 1-1969）、《乳及乳制品卫生操作规范》（CXC 57-2004）以及《卫生操作规范》和《生产操作规范》等其他相关法典文本。产品应符合《食品微生物标准制定与实施原则和准则》（CXG 21-1997）规定的所有微生物标准。

7 标识

除符合《预包装食品标识通用标准》（CXS 1-1985）和《乳品术语》（CXS 206-1999）外，还应符合下列具体规定。

7.1 产品名称

马苏里拉干酪名称应符合《预包装食品标识通用标准》（CXS 1-1985）第4.1条，只要产品符合该标准，产品销售国的习惯拼写也可被使用。

只要干酪符合本标准，就可选择使用此名称。对符合本标准但不使用此名称的干酪，应按《干酪通则》（CXS 283-1978）命名规定来命名。

水分含量高的马苏里拉干酪在命名时须附上修饰语，描述产品真实特性。

对于脂肪含量低于或高于本标准第3.3条规定的参考范围，但高于绝对最低含量的产品，在命名时须附上适当修饰语，作为名称的一部分或处于与名称同一视野内的显眼位置，说明所做调整或脂肪含量（以干物质中脂肪含量或质量百分比表示，视产品销售国的接受情况）。适当修饰语是指《干酪通则》（CXS 283-1978）第7.2条所规定的特征描述用语或是符合《营养和保健声明使用准则》（CXG 23-1997）[1]的营养声明。

该名称也可用于符合本标准规定的干酪做成的切块、切片、切碎或磨碎产品。

7.2 原产国

应当标记原产国（即生产国，而非名字起源国）。当产品在第二个国家有实质性改造[2]，则以标识目的进行改造的国家应视为原产国。

7.3 乳脂含量说明

乳脂含量应以销售国可接受的方式声明：（i）质量百分比形式；（ii）干物质中脂肪百分比形式；（iii）如果产品标识上标明了份数，也可用每份中乳脂重量（g）表示。

7.4 非零售包装标识

在包装容器上除要标明产品名称、批次、生产厂家或包装商的名称和地址外，在包装容器上或附随说明书中，也应按本标准第7条和《预包装食品标识通用标准》（CXS 1-1985）第4.1~4.8条所要求的信息以及储存说明（在必要的地方）加以陈述。如果批次、生产厂家或包装商的名称和地址可以在附随的文件标明，则可以用一个识别标识来代替。

8 抽样和分析方法

为检查产品是否符合本标准，应采用《分析和抽样推荐性方法》（CXS 234-1999）中涉及本标准规定的分析和抽样方法。

判定其他加工工艺与"帕斯塔菲拉塔"工艺是否对等：使用激光共聚焦扫描显微镜鉴定其特有结构。

---

[1] 为比较营养声明，干物质中脂肪含量最低参考值为40%。
[2] 例如，重新包装、切块、切片、切碎和磨碎不构成实质性改变。

附录

附 加 信 息

以下补充信息对前文各条款的规定不构成影响,前文规定内容对产品标识、食品名称的使用以及食品安全性至关重要。

水分含量高的马苏里拉

1 加工方法

1.1 主发酵的微生物是嗜热链球菌和乳酸菌。

1.2 水牛乳制成的产品须置于高浓度冷盐水中。

STANDARD FOR MOZZARELLA

CXS 262-2006

Adopted in 2006. Amended in 2010, 2013, 2016, 2018, 2019.

1 SCOPE

This Standard applies to Mozzarella intended for direct consumption or for further processing, in conformity with the description in Section 2 of this Standard.

2 DESCRIPTION

Mozzarella is an unripened cheese in conformity with the *General Standard for Cheese* (CXS 283-1978) and the *Standard for Unripened Cheese Including Fresh Cheese* (CXS 221-2001). It is a smooth elastic cheese with a long stranded parallel-orientated fibrous protein structure without evidence of curd granules. The cheese is rindless[1] and may be formed into various shapes.

Mozzarella with a high moisture content is a soft cheese with overlying layers that may form pockets containing liquid of milky appearance. It may be packed with or without the liquid. The cheese has a near white colour.

Mozzarella with a low moisture content is a firm/semi-hard homogeneous cheese without holes and is suitable for shredding.

Mozzarella is made by "pasta filata" processing, which consists of heating curd of a suitable pH value kneading and stretching until the curd is smooth and free from lumps. Still warm, the curd is cut and moulded, then firmed by cooling. Other processing techniques, which give end products with the same physical, chemical and organoleptic characteristics are allowed.

3 ESSENTIAL COMPOSITION AND QUALITY FACTORS

3.1 Raw materials

Cows' milk or buffaloes' milk, or their mixtures, and products obtained from these milks.

3.2 Permitted ingredients

- Starter cultures of harmless lactic acid and/or flavour producing bacteria and cultures of other harmless micro-organisms;
- Rennet or other safe and suitable coagulating enzymes;
- Sodium chloride and potassium chloride as a salt substitute;
- Safe and suitable processing aids;
- Vinegar;
- Potable water;
- Rice, corn and potato flours and starches: Notwithstanding the provisions in the *General Standard for* Cheese (CXS 283-1978), these substances can be used in the

---

[1] The cheese has been kept in such a way that no rind is developed (a "rindless" cheese).

same function as anti-caking agents for treatment of the surface of cut, sliced, and shredded Mozzarella with a low moisture content only, provided they are added only in amounts functionally necessary as governed by Good Manufacturing Practice, taking into account any use of the anti-caking agents listed in Section 4.

3.3 Composition

Milk constituent	Minimum content (m/m)	Maximum content (m/m)	Reference level (m/m)	
Milkfat in dry matter: with high moisture with low moisture	20% 18%	Not restricted Not restricted	40% to 50% 40% to 50%	
Dry matter	Depending on the fat in dry matter content, according to the table below			
	Fat in dry matter content (m/m)		Corresponding minimum dry matter content (m/m)	
			With high moisture	With low moisture
	Equal to or above 18% but less than 30%		34%	–
	Equal to or above 20% but less than 30%		–	24%
	Equal to or above 30% but less than 40%		39%	26%
	Equal to or above 40% but less than 45%		42%	29%
	Equal to or above 45% but less than 50%		45%	31%
	Equal to or above 50% but less than 60%		47%	34%
	Equal to or above 60% but less than 85%		53%	38%

Compositional modifications beyond the minima and maxima specified above for milkfat and dry matter are not considered to be in compliance with Section 4.3.3 of the *General Standard for the Use of Dairy Terms* (CXS 206-1999).

4 FOOD ADDITIVES

Only those additives classes indicated as justified in the table below may be used for the product categories specified. Within each additive class, and where permitted according to the table, only those food additives listed below may be used and only within the functions and limits specified.

Additive functional class	JUSTIFIED USE			
	Mozzarella with low moisture content		Mozzarella with high moisture content	
	Cheese mass	Surface treatment	Cheese mass	Surface treatment
Colours	×[a]	–	×[a]	–
Bleaching agents	–	–	–	–
Acidity regulators	×	–	×	–
Stabilizers	×	–	×	–
Thickeners	×	–	×	–

Additive functional class	JUSTIFIED USE			
	Mozzarella with low moisture content		Mozzarella with high moisture content	
	Cheese mass	Surface treatment	Cheese mass	Surface treatment
Emulsifiers	–	–	–	–
Antioxidants	–	–	–	–
Preservatives	×	×	×	×(c)
Foaming agents	–	–	–	–
Anticaking agents	–	×(b)	–	×(d)

(a) Only to obtain the colour characteristics described in Section 2.
(b) For the surface of sliced, cut, shredded or grated cheese only.
(c) Only for high-moisture mozzarella not packaged in liquid.
(d) For the surface treatment of shredded and/or diced cheese only.
× The use of additives belonging to the class is technologically justified.
– The use of additives belonging to the class is not technologically justified.

INS no.	Name of additive	Maximum level
Preservatives		
200	Sorbic acid	1 000 mg/kg singly or in combination as sorbic acid
202	Potassium sorbate	
203	Calcium sorbate	
234	Nisin	12.5 mg/kg
235	Natamycin (pimaricin)	Not exceeding 2 mg/dm^2 and not present in a depth of 5 mm
280	Propionic acid	Limited by GMP
281	Sodium propionate	
282	Calcium propionate	
283	Potassium propionate	
Acidity regulators		
170(i)	Calcium carbonate	Limited by GMP
260	Acetic acid, glacial	
261(i)	Potassium acetate	
261(ii)	Potassium diacetate	

INS no.	Name of additive	Maximum level
262(i)	Sodium acetate	Limited by GMP
263	Calcium acetate	
270	Lactic acid, *L-, D-* and *DL-*	
296	Malic acid, *DL-*	
325	Sodium lactate	
326	Potassium lactate	
327	Calcium lactate	
330	Citric acid	
338	Phosphoric acid	880 mg/kg as phosphorous
350(i)	Sodium hydrogen *DL*-malate	Limited by GMP
350(ii)	Sodium malate	
352(ii)	Calcium malate, *D, L-*	
500(i)	Sodium carbonate	
500(ii)	Sodium hydrogen carbonate	
500(iii)	Sodium sesquicarbonate	
501(i)	Potassium carbonate	
501(ii)	Potassium hydrogen carbonate	
504(i)	Magnesium carbonate	
504(ii)	Magnesium hydrogen carbonate	
507	Hydrochloric acid	
575	Glucono-*delta*-lactone	
577	Potassium gluconate	
578	Calcium gluconate	
Stabilizers		
331(i)	Sodium dihydrogen citrate	Limited by GMP
332(i)	Potassium dihydrogen citrate	
333	Calcium citrates	
339(i)	Sodium dihydrogen phosphate	400 mg/kg, singly or in combination, expressed as phosphorus

INS no.	Name of additive	Maximum level
339(ii)	Disodium hydrogen phosphate	400 mg/kg, singly or in combination, expressed as phosphorus
339(iii)	Trisodium phosphate	
340(i)	Potassium dihydrogen phosphate	
340(ii)	Dipotassium hydrogen phosphate	
340(iii)	Tripotassium phosphate	
341(i)	Calcium dihydrogen phosphate	
341(ii)	Calcium hydrogen phosphate	
341(iii)	Tricalcium phosphate	
342(i)	Ammonium dihydrogen phosphate	
342(ii)	Diammonium hydrogen phosphate	
343(ii)	Magnesium hydrogen phosphate	
343(iii)	Trimagnesium phosphate	
450(i)	Disodium diphosphate	
450(iii)	Tetrasodium diphosphate	
450(v)	Tetrapotassium diphosphate	
450(vi)	Dicalcium diphosphate	
451(i)	Pentasodium triphosphate	
451(ii)	Pentapotassium triphosphate	
452(i)	Sodium polyphosphate	
452(ii)	Potassium polyphosphate	
452(iv)	Calcium polyphosphate	
452(v)	Ammonium polyphosphate	
406	Agar	Limited by GMP
407	Carrageenan	
407a	Processed euchema seaweed (PES)	

INS no.	Name of additive	Maximum level
410	Carob bean gum	Limited by GMP
412	Guar gum	
413	Tragacanth gum	
415	Xanthan gum	
416	Karaya gum	
417	Tara gum	
440	Pectins	
466	Sodium carboxymethyl cellulose (Cellulose gum)	
Colours		
140	Chlorophylls	Limited by GMP
141(i)	Chlorophyll copper complexes	5 mg/kg, singly or in combination
141(ii)	Chlorophyllin copper complex, sodium and potassium salts	
171	Titanium dioxide	Limited by GMP
Anticaking agents		
460(i)	Microcrystalline cellulose (Cellulose gel)	Limited by GMP
460(ii)	Powdered cellulose	
551	Silicon dioxide, amorphous	10 000 mg/kg, singly or in combination as silicon dioxide
552	Calcium silicate	
553(i)	Magnesium silicate, synthetic	

* For the definition of cheese surface and rind see Appendix to the *General Standard for Cheese* (CXS 283-1978).

5 CONTAMINANTS

The products covered by this Standard shall comply with the Maximum Levels for contaminants that are specified for the product in the *General Standard for Contaminants and Toxins in Food and Feed* (CXS 193-1995).

The milk used in the manufacture of the products covered by this Standard shall comply with the Maximum Levels for contaminants and toxins specified for milk by the *General Standard for Contaminants and Toxins in Food and Feed* (CXS 193-1995) and with the maximum residue limits for veterinary drug residues and pesticides established for milk by the CAC.

6　HYGIENE

It is recommended that the product covered by the provisions of this standard be prepared and handled in accordance with the appropriate sections of the *General Principles of Food Hygiene* (CXC 1-1969), the *Code of Hygienic Practice for Milk and Milk Products* (CXC 57-2004) and other relevant Codex texts such as Codes of Hygienic Practice and Codes of Practice. The products should comply with any microbiological criteria established in accordance with the *Principles and Guidelines for the Establishment and Application of Microbiological Criteria Related to Foods* (CXG 21-1997).

7　LABELLING

In addition to the provisions of the *General Standard for the Labelling of Prepackaged Foods* (CXS 1-1985) and the *General Standard for the Use of Dairy Terms* (CXS 206-1999), the following specific provisions apply.

7.1　Name of the food

The name Mozzarella may be applied in accordance with Section 4.1 of the *General Standard for the Labelling of Prepackaged Foods* (CXS 1-1985), provided that the product is in conformity with this Standard. Where customary in the country of retail sale, alternative spelling may be used.

The use of the name is an option that may be chosen only if the cheese complies with this standard. Where the name is not used for a cheese that complies with this Standard, the naming provisions of the *General Standard for Cheese* (CXS 283-1978) apply.

The designation of Mozzarella with a high moisture content shall be accompanied by a qualifying term describing the true nature of the product.

The designation of products in which the fat content is below or above the reference range but above the absolute minimum specified in Section 3.3 of this Standard shall be accompanied by an appropriate qualification describing the modification made or the fat content (expressed as fat in dry matter or as percentage by mass whichever is acceptable in the country of retail sale), either as part of the name or in a prominent position in the same field of vision. Suitable qualifiers are the appropriate characterizing terms specified in Section 7.2 of the *General Standard for Cheese* (CXS 283-1978) or a nutritional claim in accordance with the *Guidelines for the Use of Nutritional Claims* (CXG 23-1997)[1].

The designation may also be used for cut, sliced, shredded or grated products made from cheese which cheese is in conformity with this Standard.

7.2　Country of origin

The country of origin (which means the country of manufacture, not the country in which the name originated) shall be declared. When the product undergoes substantial transformation[2] in a second country, the country in which the transformation is performed shall be considered to be the country of origin for the purpose of labelling.

1　For the purpose of comparative nutritional claims, the minimum fat content of 40% fat in dry matter constitutes the references.

2　For instance, repackaging, cutting, slicing, shredding and grating is not regarded as substantial transformation.

7.3 Declaration of milkfat content

The milk fat content shall be declared in a manner found acceptable in the country of retail sale, either (i) as a percentage by mass, (ii) as a percentage of fat in dry matter, or (iii) in grams per serving as quantified in the label, provided that the number of servings is stated.

7.4 Labelling of non retail containers

Information specified in Section 7 of this Standard and Sections 4.1 to 4.8 of the *General Standard for the Labelling of Prepackaged Foods* (CXS 1-1985) and, if necessary, storage instructions, shall be given either on the container or in accompanying documents, except that the name of the product, lot identification, and the name of the manufacturer or packer shall appear on the container, and in the absence of such a container, on the product itself. However, lot identification and the name and address may be replaced by an identification mark, provided that such mark is clearly identifiable with the accompanying documents.

8 METHODS OF SAMPLING AND ANALYSIS

For checking the compliance with this standard, the methods of analysis and sampling contained in the *Recommended Methods of Analysis and Sampling* (CXS 234-1999) relevant to the provisions in this Standard, shall be used.

Determination of equivalency between "pasta filata" processing and other processing techniques: Identification of the typical structure by confocal laser scanning microscopy.

APPENDIX

ADDITIONAL INFORMATION

The additional information below does not affect the provisions in the preceding sections which are those that are essential to the product identity, the use of the name of the food and the safety of the food.

Mozzarella with a high moisture content

1 **Method of manufacture**

1.1 The principal starter culture micro-organisms are *Streptococcus thermophilus* and/or *Lactococcus* spp.

1.2 Products made from buffalo's milk shall be salted in cold brine.

切达干酪

STANDARD FOR CHEDDAR

CXS 263-1966

原为CODEX STAN C-1-1966。1966年通过。

2007年修订，2008年、2010年、2013年、2018年、2019年修正。

1 范围

本标准适用于符合本标准第2条所述即食或用于进一步加工的切达干酪。

2 说明

切达干酪是指符合《干酪通则》（CXS 283-1978）的成熟硬质干酪。干酪外观呈白色、银白色、淡黄色、橘黄色，其质地坚硬（用拇指摁压时），表面光滑似蜡。无气孔，允许个别开口或裂缝。干酪加工和销售去或不去外皮均可[1]。

对于即食的切达干酪，调制味道和外观特征的催熟程序在7～15 ℃下通常需要5周，取决于所需成熟度。在干酪具备和前述催熟程序同样的物理、生化和感官特性的情况下，可以使用其他催熟条件（包括添加加速熟化的酶）。如有合理技术和/或贸易需求，用进一步加工的切达干酪则无需具备同等成熟度。

3 基本成分和质量要求

3.1 基本原料

牛乳、水牛乳或它们的混合乳，以及源于这些乳的制品。

3.2 其他配料

- 无害乳酸发酵剂和/或产香菌和其他无害微生物的培养剂；
- 凝乳酶或其他安全适宜的凝固酶；
- 作为代盐制品的氯化钠和氯化钾；
- 饮用水；
- 促进成熟的安全、适宜的酶；
- 安全、合适的加工助剂；
- 大米、玉米、马铃薯粉和淀粉：尽管有《干酪通则》（CXS 283-1978），但这些物质同样可作为抗结块剂处理切块、切片和切碎产品表面，只要添加量是

---

[1] 并非在销售前去掉外皮，相反干酪在成熟后，以不产生外皮方式保存（无皮干酪）。在无皮干酪加工过程中，使用催熟膜。催熟膜也可以成为保护干酪的外衣。无皮干酪也可参见《干酪通则》附件（CXS 283-1978）。

良好操作规范所要求达到期待功能的必需量，同时考虑第4条规定的抗结块剂的使用。

3.3 成分

乳成分	最低含量（m/m）	最高含量（m/m）	参考含量（m/m）
干物质乳脂	22%	/	48%~60%
干物质	根据干物质脂肪含量，参见下表		
	干物质脂肪含量（m/m）	相应最低干物质含量（m/m）	
	≥22%，<30%	49%	
	≥30%，<40%	53%	
	≥40%，<48%	57%	
	≥48%，<60%	61%	
	≥60%	66%	

乳脂和干物质含量超过规定的最大值和最小值被视为不符合《乳品术语》（CXS 206-1999）第4.3.3条。

4 食品添加剂

只有在下表中标明的添加剂类别，可用于指定产品类别。根据《食品添加剂通用标准》（CXS 192-1995）表1和表2在食品类别01.6.2.1（成熟干酪，包括外皮）中使用的抗结块剂、着色剂和防腐剂，以及表3中仅部分酸度调节剂和抗结块剂可用于符合本标准的食品。

添加剂功能类别	合理使用	
	干酪量	表面/外皮处理
着色剂	×[a]	−
漂白剂	−	−
酸度调节剂	×	−
稳定剂	−	−
增稠剂	−	−
乳化剂	−	−
抗氧化剂	−	−
防腐剂	×	×

添加剂功能类别	合理使用	
	干酪量	表面/外皮处理
发泡剂	-	-
抗结块剂	-	×(b)

（a）仅为获得第2条所述颜色特征。
（b）仅用于切片、切块、切碎或磨碎的乳酪表面。
× 使用该类别添加剂符合技术要求。
- 使用该类别添加剂不符合技术要求。

5　污染物

本标准所涉及的产品应符合《食品及饲料中污染物和毒素通用标准》（CXS 193-1995）规定的污染物最大限量。

本标准所涉及的产品在加工中使用的牛乳应符合《食品及饲料中污染物和毒素通用标准》（CXS 193-1995）规定的乳中污染物和毒素最大限量以及国际食品法典委员会设定的农药和兽药最大残留限量。

6　卫生要求

建议本标准所涉及产品应遵循《食品卫生总则》（CXC 1-1969）、《乳及乳制品卫生操作规范》（CXC 57-2004）以及《卫生操作规范》和《生产操作规范》等其他相关法典文本。产品应符合《食品微生物标准制定与实施原则和准则》（CXG 21-1997）规定的所有微生物标准。

7　标识

除符合《预包装食品标识通用标准》（CXS 1-1985）和《乳品术语》（CXS 206-1999）外，还应符合下列具体规定。

7.1　产品名称

切达干酪名称应符合《预包装食品标识通用标准》（CXS 1-1985）4.1条，只要产品符合该标准，产品销售国的习惯拼写也可被使用。

只要干酪符合本标准，就可选择使用此名称。对符合本标准但不使用此名称的干酪，应按《干酪通则》（CXS 283-1978）命名规定来命名。

对于脂肪含量超过本标准第3.3条规定的参考含量范围，但高于绝对最低含量的产品，在命名时须附上适当修饰语，作为名称的一部分或处于与名称同一视野内的显眼位置，说明所做调整或脂肪含量（以干物质中脂肪含量或质量百分比表示，视产品销售国的接受情况）。适当修饰语是指《干酪通则》（CXS 283-1978）第7.2条所规定的特征描述用语或是符合《营养和保健声明使用准则》（CXG 23-1997）[1]的营养声明。

该名称也可用于符合本标准规定的干酪做成的切块、切片、切碎或磨碎产品。

7.2 原产国

应当标记原产国（即生产国，而非名称起源国）。当产品在第二个国家有实质性改造[2]，则以标识目的进行改造的国家应视为原产国。

7.3 乳脂含量说明

乳脂含量应以销售国可接受的方式声明：（i）质量百分比形式；（ii）干物质中脂肪百分比形式；（iii）如果产品标识上标明了份数，也可用每份中乳脂重量（g）表示。

7.4 日期标识

尽管《预包装食品标识通用标准》（CXS 1-1985）第4.7.1条规定，仍应说明生产日期而非最低保存时间信息，只要产品并非直接出售给最终消费者。

7.5 非零售包装标识

在包装容器上除要标明产品名称、批次、生产厂家或包装商的名称和地址外，在包装容器上或附随说明书中，也应按本标准第7条和《预包装食品标识通用标准》（CXS 1-1985）第4.1～4.8条所要求的信息以及储存说明（在必要的地方）加以陈述。如果批次、生产厂家或包装商的名称和地址可以在附随的文件标明，则可以用一个识别标识来代替。

8 抽样和分析方法

为检查产品是否符合本标准，应采用《分析和抽样推荐性方法》（CXS 234-1999）中涉及本标准规定的分析和抽样方法。

---

1 为比较营养声明，干物质中脂肪含量最低参考值为48%。
2 例如，重新包装，切块、切片、切碎和磨碎不构成实质性改变。

附录

附 加 信 息

以下补充信息对前文各条款的规定不构成影响,前文规定内容对产品标识、食品名称的使用以及食品安全性至关重要。

1 加工方法

1.1 发酵剂由不产气的乳酸菌组成。

1.2 凝固后,凝乳被分割,且凝乳在其乳清中加热至凝固温度以上。将凝乳从乳清中分离出来,并搅拌或形成切达干酪。在传统生产中,凝乳被切成块,并逐渐堆积,保持凝乳温热,从而导致凝乳变得可压缩、光滑且富有弹性。做切达干酪后凝乳被研磨。当达到所需的酸度盐腌凝乳,凝乳和盐混合并成形。若其他加工技术使最终产品具有相同物理,生化和感官特性,也可被应用。

STANDARD FOR CHEDDAR

CXS 263-1966

Formerly CODEX STAN C-1-1966. Adopted in 1996. Revised in 2007.

Amended in 2008, 2010, 2013, 2018, 2019.

1 SCOPE

This Standard applies to Cheddar intended for direct consumption or for further processing in conformity with the description in Section 2 of this Standard.

2 DESCRIPTION

Cheddar is a ripened hard cheese in conformity with the *General Standard for Cheese* (CXS 283-1978). The body has a near white or ivory through to light yellow or orange colour and a firm-textured (when pressed by thumb), smooth and waxy texture. Gas holes are absent, but a few openings and splits are acceptable. The cheese is manufactured and sold with or without[1] rind which may be coated.

For Cheddar ready for consumption, the ripening procedure to develop flavour and body characteristics is normally from 5 weeks at 7–15 °C depending on the extent of maturity required. Alternative ripening conditions (including the addition of ripening enhancing enzymes)may be used, provided the cheese exhibits similar physical, biochemical and sensory properties as those achieved by the previously stated ripening procedure. Cheddar intended for further processing need not exhibit the same extent of ripening when justified through technical and/or trade needs.

3 ESSENTIAL COMPOSITION AND QUALITY FACTORS

3.1 Raw materials

Cows' milk or buffaloes' milk, or their mixtures, and products obtained from these milks.

3.2 Permitted ingredients

- Starter cultures of harmless lactic acid and/or flavour producing bacteria and cultures of other harmless microorganisms;
- Rennet or other safe and suitable coagulating enzymes;
- Sodium chloride and potassium chloride as a salt substitute;
- Potable water;
- Safe and suitable enzymes to enhance the ripening process;
- Safe and suitable processing aids;
- Rice, corn and potato flours and starches: Notwithstanding the provisions in the

---

[1] This is not to mean that the rind has been removed before sale, instead the cheese has been ripened and/or kept in such a way that no rind is developed (a "rindless" cheese). Ripening film is used in the manufacture of rindless cheese. Ripening film may also constitute the coating that protects the cheese. For rindless cheese see also the Appendix to the *General Standard for Cheese* (CXS 283-1978).

General Standard for Cheese (CXS 283-1978), these substances can be used in the same function as anti-caking agents for treatment of the surface of cut, sliced, and shredded products only, provided they are added only in amounts functionally necessary as governed by Good Manufacturing Practice, taking into account any use of the anti-caking agents listed in section 4.

3.3 Composition

Milk constituent	Minimum content (m/m)	Maximum content (m/m)	Reference level (m/m)
Milk fat in dry matter	22%	Not restricted	48% to 60%
Dry matter	Depending on the fat in dry matter content, according to the table below		
	Fat in dry matter content (m/m)		**Corresponding minimum dry matter content** (m/m)
	Equal to or above 22% but less than 30%		49%
	Equal to or above 30% but less than 40%		53%
	Equal to or above 40% but less than 48%		57%
	Equal to or above 48% but less than 60%		61%
	Equal to or above 60%		66%

Compositional modifications beyond the minima and maxima specified above for milk fat and dry matter are not considered to be in compliance with section 4.3.3 of the *General Standard for the Use of Dairy Terms* (CXS 206-1999).

4 FOOD ADDITIVES

Only those additives classes indicated as justified in the table below may be used for the product categories specified. Anticaking agents, colours and preservatives used in accordance with Tables 1 and 2 of the *General Standard for Food Additives* (CXS 192-1995) in food category 01.6.2.1 (Ripened cheese, includes rind) and only certain acidity regulators, anticaking agents and colours in Table 3 are acceptable for use in foods conforming to this standard.

Additive functional class	Justified use	
	Cheese mass	**Surface/rind treatment**
Colours	×[a]	–
Bleaching agents	–	–
Acidity regulators	×	–
Stabilizers	–	–
Thickeners	–	–
Emulsifiers	–	–
Antioxidants	–	–

Additive functional class	Justified use	
	Cheese mass	Surface/rind treatment
Preservatives	×	×
Foaming agents	–	–
Anti–caking agents	–	×(b)

(a) Only to obtain the colour characteristics, as described in Section 2.
(b) For the surface of sliced, cut, shredded or grated cheese, only.
× The use of additives belonging to the class is technologically justified.
– The use of additives belonging to the class is not technologically justified.

5 CONTAMINANTS

The products covered by this Standard shall comply with the Maximum Levels for contaminants that are specified for the product in the *General Standard for Contaminants and Toxins in Food and Feed* (CXS 193-1995).

The milk used in the manufacture of the products covered by this Standard shall comply with the Maximum Levels for contaminants and toxins specified for milk by the *General Standard for Contaminants and Toxins in Food and Feed* (CXS 193-1995) and with the maximum residue limits for veterinary drug residues and pesticides established for milk by the CAC.

6 HYGIENE

It is recommended that the product covered by the provisions of this standard be prepared and handled in accordance with the appropriate sections of the *General Principles of Food Hygiene* (CXC 1-1969), the *Code of Hygienic Practice for Milk and Milk Products* (CXC 57-2004) and other relevant Codex texts such as Codes of Hygienic Practice and Codes of Practice. The products should comply with any microbiological criteria established in accordance with the *Principles and Guidelines for the Establishment and Application of Microbiological Criteria Related to Foods* (CXG 21-1997).

7 LABELLING

In addition to the provisions of the *General Standard for the Labelling of Prepackaged Foods* (CXS 1-1985) and the *General Standard for the Use of Dairy Terms* (CXS 206-1999), the following specific provisions apply.

7.1 Name of the food

The name Cheddar may be applied in accordance with Section 4.1 of the *General Standard for the Labelling of Prepackaged Foods* (CXS 1-1985), provided that the product is in conformity with this Standard. Where customary in the country of retail sale, alternative spelling may be used.

The use of the name is an option that may be chosen only if the cheese complies with this standard. Where the name is not used for a cheese that complies with this standard, the naming provisions of the *General Standard for Cheese* (CXS 283-1978) apply.

The designation of products in which the fat content is below or above the reference

range but above the absolute minimum specified in section 3.3 of this Standard shall be accompanied by an appropriate qualification describing the modification made or the fat content (expressed as fat in dry matter or as percentage by mass, whichever is acceptable in the country of retail sale), either as part of the name or in a prominent position in the same field of vision. Suitable qualifiers are the appropriate characterizing terms specified in Section 7.2 of the *General Standard for Cheese* (CXS 283-1978) or a nutritional claim in accordance with the *Guidelines for Use of Nutritional and Health Claims* (CXG 23-1997)[1].

The designation may also be used for cut, sliced, shredded or grated products made from cheese which cheese is in conformity with this Standard.

7.2 Country of origin

The country of origin (which means the country of manufacture, not the country in which the name originated) shall be declared. When the product undergoes substantial transformation[2] in a second country, the country in which the transformation is performed shall be considered to be the country of origin for the purpose of labelling.

7.3 Declaration of milk fat content

The milk fat content shall be declared in a manner found acceptable in the country of retail sale either (i) as a percentage by mass, (ii) as a percentage of fat in dry matter, or (iii) in grams per serving as quantified in the label, provided that the number of servings is stated.

7.4 Date marking

Notwithstanding the provisions of Section 4.7.1 of the *General Standard for the Labelling of Prepackaged Foods* (CXS 1-1985), the date of manufacture may be declared instead of the minimum durability information, provided that the product is not intended to be purchased as such by the final consumer.

7.5 Labelling of non-retail containers

Information specified in Section 7 of this Standard and Sections 4.1 to 4.8 of the *General Standard for the Labelling of Prepackaged Foods* (CXS 1-1985) and, if necessary, storage instructions, shall be given either on the container or in accompanying documents, except that the name of the product, lot identification, and the name of the manufacturer or packer shall appear on the container, and in the absence of such a container, on the product itself. However, lot identification and the name and address may be replaced by an identification mark, provided that such mark is clearly identifiable with the accompanying documents.

8 METHODS OF SAMPLING AND ANALYSIS

For checking the compliance with this standard, the methods of analysis and sampling contained in the *Recommended Methods of Analysis and Sampling* (CXS 234-1999) relevant to the provisions in this standard, shall be used.

---

1 For the purpose of comparative nutritional claims, the minimum fat content of 48% fat in dry matter constitutes the reference.

2 For instance, repackaging, cutting, slicing, shredding and grating is not regarded as substantial transformation.

APPENDIX

ADDITIONAL INFORMATION

The additional information below does not affect the provisions in the preceding sections which are those that are essential to the product identity, the use of the name of the food and the safety of the food.

1 **Method of manufacture**

1.1 Starter cultures consist of non-gas forming lactic acid producing bacteria.

1.2 After coagulation, the curd is cut and heated in its whey to a temperature above the coagulation temperature. The curd is separated from the whey and stirred or cheddared. In traditional manufacture the curd is cut into blocks which are turned and progressively piled, keeping the curd warm, which results in the curd becoming compressed, smooth and elastic. After cheddaring the curd is milled. When the desired acidity is reached the curd is salted. The curd and salt are then mixed and moulded. Other processing techniques, which give end products with the same physical, chemical and organoleptic characteristics may be applied.

旦伯干酪

STANDARD FOR DANBO

CXS 264-1966

原为CODEX STAN C-3-1966。1966年通过。2007年修订。
2008年、2010年、2013年、2018年、2019年修正。

1 范围

本标准适用于符合本标准第2条所述即食或用于进一步加工的旦伯干酪。

2 说明

旦伯干酪是指符合《干酪通则》（CXS 283-1978）成熟的硬质或半硬质干酪。干酪外观呈白色、银白色、淡黄色或橘黄色，其质地坚硬（用拇指摁压）、适合切割，有少数或多个均匀分布、光滑、似圆豆大小的气孔（大多数直径为10 mm），允许个别开口和裂缝。外形为扁方形或平行六面形。干酪制造和销售有无坚硬或潮湿外皮均可[1]，干酪可涂抹。

对于即食的旦伯干酪，调制味道和外观特征的催熟程序在12~20 ℃下通常需要3周，取决于所需成熟度。在干酪具备和前述催熟程序同样的物理、生化和感官特性的情况下，可以使用其他催熟条件（包括添加加速熟化的酶）。如有合理技术或贸易需求，用于进一步加工的旦伯干酪无需具有同等成熟度。

3 基本成分和质量要求

3.1 基本原料

牛乳、水牛乳或它们的混合乳，以及源于这些乳的制品。

3.2 其他配料

- 无害乳酸发酵剂和/或产香菌和其他无害微生物的培养剂；
- 凝乳酶或其他安全、适宜的凝固酶；
- 作为代盐制品的氯化钠和氯化钾；
- 饮用水；
- 促进成熟的安全、适宜的酶；

---

1 并非在销售前去掉外皮，相反干酪在成熟后，以不产生外皮方式保存（无皮干酪）。在无皮干酪加工过程中，使用催熟膜。催熟膜也可以成为保护干酪的外衣。无皮干酪也可参见《干酪通则》附件（CXS 283-1978）。

- 安全、适宜的加工助剂；
- 大米、玉米、马铃薯粉和淀粉：尽管有《干酪通则》（CXS 283-1978），但这些物质同样可作为抗结块剂处理切块、切片和切碎产品表面，只要添加量是良好操作规范所要求达到期待功能的必需量，同时考虑第4条规定的抗结块剂的使用。

3.3 成分

乳成分	最低含量（m/m）	最高含量（m/m）	参考含量（m/m）
干物质乳脂	20%	/	45%~55%

干物质	根据干物质脂肪含量，参见下表	
	干物质脂肪含量（m/m）	相应最低干物质含量（m/m）
	≥20%，<30%	41%
	≥30%，<40%	44%
	≥40%，<45%	50%
	≥45%，<55%	52%
	≥55%	57%

乳脂和干物质含量超过规定的最大值和最小值被视为不符合《乳品术语》（CXS 206-1999）第4.3.3条。

4 食品添加剂

只有下表中标注为合理使用的添加剂类别才可以用于指定产品类别。根据《食品添加剂通用标准》（CXS 192-1995）表1和表2在食品类别01.6.2.1（成熟干酪，包括外皮）中使用的抗结块剂、着色剂和防腐剂，以及表3中仅部分酸度调节剂和抗结块剂可用于符合本标准的食品。

添加剂功能类别	合理使用	
	干酪块	表面/外皮处理
着色剂	×(a)	—
漂白剂	—	—
酸度调节剂	×	—

添加剂功能类别	合理使用	
	干酪块	表面/外皮处理
稳定剂	–	–
增稠剂	–	–
乳化剂	–	–
抗氧化剂	–	–
防腐剂	×	×
发泡剂	–	–
抗结块剂	–	×[b]

（a）仅为获得第2条所述颜色特征。
（b）仅用于切片、切块、切碎或磨碎的乳酪表面。
× 使用该类别添加剂符合技术要求。
– 使用该类别添加剂不符合技术要求。

5 污染物

本标准所涉及的产品应符合《食品及饲料中污染物和毒素通用标准》（CXS 193-1995）规定的污染物最大限量。

本标准所涉及的产品在加工中使用的牛乳应符合《食品及饲料中污染物和毒素通用标准》（CXS 193-1995）规定的乳中污染物和毒素最大限量以及国际食品法典委员会设定的农药和兽药最大残留限量。

6 卫生要求

建议本标准所涉及的产品应遵循《食品卫生总则》（CXC 1-1969）、《乳及乳制品卫生操作规范》（CXC 57-2004）和其他相关法典文本，如《卫生操作规范》和《生产操作规范》的规定。产品应符合《食品微生物标准制定与实施原则和准则》（CXG 21-1997）规定的所有微生物标准。

7 标识

除符合《预包装食品标识通用标准》（CXS 1-1985）和《乳品术语》（CXS 206-1999）外，还应符合下列具体规定。

7.1 产品名称

旦伯干酪名称应符合《预包装食品标识通用标准》（CXS 1-1985）第4.1条，只要产品符合该标准，产品销售国的习惯拼写也可被使用。

只要干酪符合本标准，就可选择使用此名称。对符合本标准但不使用此名称的干酪，应按《干酪通则》（CXS 283-1978）命名规定来命名。

对于脂肪含量超过本标准第3.3条规定的参考含量范围，但高于绝对最低含量的产品，在命名时须附上适当修饰语，作为名称的一部分或处于与名称同一视野内的显眼位置，说明所做调整或脂肪含量（以干物质中脂肪含量或质量百分比表示，视产品销售国的接受情况）。适当修饰语是指《干酪通则》（CXS 283-1978）第7.2条所规定的特征描述用语或是符合《营养和保健声明使用准则》（CXG 23-1997）[1]的营养声明。

该名称也可用于符合本标准干酪做成的切块、切片、切碎或磨碎产品。

7.2 原产国

应当标记原产国（即生产国，而非名称起源国）。当产品在另一个国家发生实质性转变[2]，则以标识目的进行改进的国家应视为原产国。

7.3 乳脂含量说明

乳脂含量应以销售国可接受的方式声明：（i）质量百分比形式；（ii）干物质中脂肪百分比形式；（iii）如果产品标识上标明了份数，也可用每份中乳脂重量（g）表示。

7.4 日期标识

尽管《预包装食品标识通用标准》（CXS 1-1985）第4.7.1条规定，仍应说明生产日期而非最低保存时间信息，只要产品并非直接出售给最终消费者。

7.5 非零售包装标识

在包装容器上除要标明产品名称、批次、生产厂家或包装商的名称和地址外，在包装容器上或附随说明书中，也应按本标准第7条和《预包装食品标识通用标准》（CXS 1-1985）第4.1 ~ 4.8条所要求的信息以及储存说明（在必要的地方）加以陈述。如果批次、生产厂家或包装商的名称和地址可以在附随的文件标明，则可以用一个识别标识来代替。

8 抽样和分析方法

为检查产品是否符合本标准，应采用《分析和抽样推荐性方法》（CXS 234-1999）中涉及本标准规定的分析和抽样方法。

---

1 为比较营养声明，干物质中脂肪含量最低参考值为45%。
2 例如，重新包装，切块、切片、切碎和磨碎不构成实质性改变。

STANDARD FOR DANBO

CXS 264-1966

Formerly CODEX STAN C-3-1966. Adopted in 1966. Revised in 2007.

Amended in 2008, 2010, 2013, 2018, 2019.

1 SCOPE

This Standard applies to Danbo intended for direct consumption or for further processing in conformity with the description in Section 2 of this Standard.

2 DESCRIPTION

Danbo is a ripened firm/semi-hard cheese in conformity with the *General Standard for Cheese* (CXS 283-1978). The body has a near white or ivory through to light yellow or yellow colour and a firm-textured (when pressed by thumb) texture, suitable for cutting, with few to plentiful, evenly distributed, smooth and round pea sized (or mostly up to 10 mm in diameter) gas holes, but a few openings and splits are acceptable. The shape is flat squared or parallelepiped. The cheese is manufactured and sold with or without[1] hard or slightly moist smear ripened rind, which may be coated.

For Danbo ready for consumption, the ripening procedure to develop flavour and body characteristics is normally from 3 weeks at 12–20 °C depending on the extent of maturity required. Alternative ripening conditions (including the addition of ripening enhancing enzymes) may be used, provided the cheese exhibits similar physical, biochemical and sensory properties as those achieved by the previously stated ripening procedure. Danbo intended for further processing need not exhibit the same extent of ripening when justified through technical and/or trade needs.

3 ESSENTIAL COMPOSITION AND QUALITY FACTORS

3.1 Raw materials

Cows' milk or buffaloes' milk, or their mixtures, and products obtained from these milks.

3.2 Permitted ingredients

- Starter cultures of harmless lactic acid and/or flavour producing bacteria and cultures of other harmless micro-organisms;
- Rennet or other safe and suitable coagulating enzymes;
- Sodium chloride and potassium chloride as a salt substitute;
- Potable water;
- Safe and suitable enzymes to enhance the ripening process;
- Safe and suitable processing aids;

---

[1] This is not to mean that the rind has been removed before sale, instead the cheese has been ripened and/or kept in such a way that no rind is developed (a "rindless" cheese). Ripening film is used in the manufacture of rindless cheese. Ripening film may also constitute the coating that protects the cheese. For rindless cheese, see also the Appendix to the *General Standard for Cheese* (CXS 283-1978).

- Rice, corn and potato flours and starches: Notwithstanding the provisions in the *General Standard for Cheese* (CXS 283-1978), these substances can be used in the same function as anti-caking agents for treatment of the surface of cut, sliced, and shredded products only, provided they are added only in amounts functionally necessary as governed by Good Manufacturing Practice, taking into account any use of the anti-caking agents listed in Section 4.

3.3 Composition

Milk constituent	Minimum content (m/m)	Maximum content (m/m)	Reference level (m/m)
Milkfat in dry matter	20%	Not restricted	45% to 55%
Dry matter	Depending on the fat in dry matter content, according to the table below		
	Fat in dry matter content (m/m)		Corresponding minimum dry matter content (m/m)
	Equal to or above 20% but less than 30%		41%
	Equal to or above 30% but less than 40%		44%
	Equal to or above 40% but less than 45%		50%
	Equal to or above 45 but less than 55%		52%
	Equal to or above 55%		57%

Compositional modifications beyond the minima and maxima specified above for milk fat and dry matter are not considered to be in compliance with section 4.3.3 of the *General Standard for the Use of Dairy Terms* (CXS 206-1999).

4 FOOD ADDITIVES

Only those additives classes indicated as justified in the table below may be used for the product categories specified. Anticaking agents, colours and preservatives used in accordance with Tables 1 and 2 of the *General Standard for Food Additives* (CXS 192-1995) in food category 01.6.2.1 (Ripened cheese, includes rind) and only certain acidity regulators, anticaking agents and colours in Table 3 are acceptable for use in foods conforming to this standard.

Additive functional class	Justified use	
	Cheese mass	Surface/rind treatment
Colours	×[a]	–
Bleaching agents	–	–
Acidity regulators	×	–
Stabilizers	–	–
Thickeners	–	–
Emulsifiers	–	–

STANDARD FOR DANBO

Additive functional class	Justified use	
	Cheese mass	Surface/rind treatment
Antioxidants	–	–
Preservatives	×	×
Foaming agents	–	–
Anti–caking agents	–	×(b)

(a) Only to obtain the colour characteristics, as described in Section 2.

(b) For the surface of sliced, cut, shredded or grated cheese, only.

× The use of additives belonging to the class is technologically justified.

– The use of additives belonging to the class is not technologically justified.

5 CONTAMINANTS

The products covered by this Standard shall comply with the Maximum Levels for contaminants that are specified for the product in the *General Standard for Contaminants and Toxins in Food and Feed* (CXS 193-1995).

The milk used in the manufacture of the products covered by this Standard shall comply with the Maximum Levels for contaminants and toxins specified for milk by the *General Standard for Contaminants and Toxins in Food and Feed* (CXS 193-1995) and with the maximum residue limits for veterinary drug residues and pesticides established for milk by the CAC.

6 HYGIENE

It is recommended that the product covered by the provisions of this standard be prepared and handled in accordance with the appropriate sections of the *General Principles of Food Hygiene* (CXC 1-1969), the *Code of Hygienic Practice for Milk and Milk Products* (CXC 57-2004) and other relevant Codex texts such as Codes of Hygienic Practice and Codes of Practice. The products should comply with any microbiological criteria established in accordance with the *Principles and Guidelines for the Establishment and Application of Microbiological Criteria Related to Foods* (CXG 21-1997).

7 LABELLING

In addition to the provisions of the *General Standard for the Labelling of Prepackaged Foods* (CODEX CXS 1-1985) and the *General Standard for the Use of Dairy Terms* (CXS 206-1999), the following specific provisions apply.

7.1 Name of the food

The name Danbo may be applied in accordance with section 4.1 of the *General Standard for the Labelling of Prepackaged Foods* (CXS 1-1985), provided that the product is in conformity with this Standard. Where customary in the country of retail sale, alternative spelling may be used.

The use of the name is an option that may be chosen only if the cheese complies with this standard. Where the name is not used for a cheese that complies with this standard, the naming provisions of the *General Standard for Cheese* (CXS 283-1978) apply.

The designation of products in which the fat content is below or above the reference range but above the absolute minimum specified in section 3.3 of this Standard shall be accompanied by an appropriate qualification describing the modification made or the fat content (expressed as fat in dry matter or as percentage by mass whichever is acceptable in the country of retail sale), either as part of the name or in a prominent position in the same field of vision. Suitable qualifiers are the appropriate characterizing terms specified in Section 7.2 of the *General Standard for Cheese* (CXS 283-1978) or a nutritional claim in accordance with the *Guidelines for Use of Nutrition and Health Claims* (CXG 23-1997)[1].

The designation may also be used for cut, sliced, shredded or grated products made from cheese which cheese is in conformity with this Standard.

7.2 Country of origin

The country of origin (which means the country of manufacture, not the country in which the name originated) shall be declared. When the product undergoes substantial transformation[2] in a second country, the country in which the transformation is performed shall be considered to be the country of origin for the purpose of labelling.

7.3 Declaration of milk fat content

The milk fat content shall be declared in a manner found acceptable in the country of retail sale, either (i) as a percentage by mass, (ii) as a percentage of fat in dry matter, or (iii) in grams per serving as quantified in the label, provided that the number of servings is stated.

7.4 Date marking

Notwithstanding the provisions of Section 4.7.1 of the *General Standard for the Labelling of Prepackaged Foods* (CXS 1-1985), the date of manufacture may be declared instead of the minimum durability information, provided that the product is not intended to be purchased as such by the final consumer.

7.5 Labelling of non-retail containers

Information specified in Section 7 of this Standard and Sections 4.1 to 4.8 of the *General Standard for the Labelling of Prepackaged Foods* (CXS 1-1985) and, if necessary, storage instructions, shall be given either on the container or in accompanying documents, except that the name of the product, lot identification, and the name of the manufacturer or packer shall appear on the container, and in the absence of such a container, on the product itself. However, lot identification and the name and address may be replaced by an identification mark, provided that such mark is clearly identifiable with the accompanying documents.

8 METHODS OF SAMPLING AND ANALYSIS

For checking the compliance with this standard, the methods of analysis and sampling contained in the *Recommended Methods of Analysis and Sampling* (CXS 234-1999) relevant to the provisions in this standard, shall be used.

---

1 For the purpose of comparative nutritional claims, the minimum fat content of 45% fat in dry matter constitutes the reference.

2 For instance, repackaging, cutting, slicing, shredding and grating is not regarded as substantial transformation.

艾丹姆干酪

STANDARD FOR EDAM

CXS 265-1966

原为CODEX STAN C-4-1966。1966年通过，2007年修订。

2008年、2010年、2013年、2018年、2019年修正

1 范围

本标准适用于符合本标准第2条所述即食或用于进一步加工的艾丹姆干酪。

2 说明

艾丹姆干酪是指符合《干酪通则》（CXS 283-1978）成熟的硬质或半硬质的干酪。干酪外形呈球状、扁块状或面包状，外观呈白色、银白色、淡黄色、橘黄色，其质地坚硬（用拇指摁压）、适合切割，有少量均匀分布的米粒至豌豆大小（大多数直径不超过10 mm）的圆形气孔，其中少量开口或裂口。干酪加工及销售时带有干皮，也可以加以包装。扁块状或面包状的艾丹姆干酪也可不带皮销售[1]。对于即食的艾丹姆干酪，调制味道和外观特征的催熟程序在10～18 ℃下通常需要3周，取决于所需成熟度。在干酪具备和前述催熟程序同样的物理、生化和感官特性的情况下，可以使用其他催熟条件（包括添加加速熟化的酶）。如有合理技术或贸易需求，用于进一步加工的艾丹姆干酪无需具备同等成熟度。

3 基本成分和质量要求

3.1 基本原料

牛乳、水牛乳或它们的混合乳，以及源于这些乳的制品。

3.2 其他配料

- 无害乳酸发酵剂和/或产香菌和其他无害微生物的培养剂；
- 凝乳酶或其他安全、适宜的凝固酶；
- 作为代盐制品的氯化钠和氯化钾；
- 饮用水；
- 促进成熟的安全、适宜的酶；

---

1 并非在销售前去掉外皮，相反干酪催熟后，以不产生外皮方式保存（无皮干酪）。在无皮干酪加工过程中，使用催熟膜。催熟膜也可以成为保护干酪的外衣。无皮干酪也可参见《干酪通则》（CXS 283-1978）附件。

- 安全、适宜的加工助剂；
- 大米、玉米、马铃薯粉和淀粉：尽管有《干酪通则》（CXS 283-1978），但这些物质同样可作为抗结块剂处理切块、切片和切碎产品表面，只要添加量是良好操作规范所要求达到期待功能的必需量，同时考虑第4条规定的抗结块剂的使用。

3.3 成分

乳成分	最低含量（m/m）	最高含量（m/m）	参考含量（m/m）
干物质乳脂	30%	/	40%~50%
干物质	根据干物质脂肪含量，参见下表		

干物质脂肪含量（m/m）	相应最低干物质含量（m/m）
≥30%，<40%	47%
≥40%，<45%	51%
≥45%，<50%	55%
≥50%，<60%	57%
≥60%	62%

乳脂和干物质含量超过上述规定的最大值和最小值被视为不符合《乳品术语》（CXS 206-1999）第4.3.3条。

4 食品添加剂

只有下表中标注为合理使用的添加剂类别才可用于指定产品类别。根据《食品添加剂通用标准》（CXS 192-1995）表1和表2在食品类别01.6.2.1（成熟干酪，包括外皮）中使用的抗结块剂、着色剂和防腐剂，以及表3中仅部分酸度调节剂和抗结块剂可用于符合本标准的食品。

添加剂功能类别	合理使用	
	干酪量	表面/外皮处理
着色剂	×[a]	—
漂白剂	—	—
酸度调节剂	×	—
稳定剂	—	—
增稠剂	—	—

添加剂功能类别	合理使用	
	干酪量	表面/外皮处理
乳化剂	-	-
抗氧化剂	-	-
防腐剂	×	×
发泡剂	-	-
抗结块剂	-	×(b)

（a）仅为获得第2条所述颜色特征。
（b）仅用于切片、切块、切碎或磨碎的乳酪表面。
　×　使用该类别添加剂符合技术要求。
　-　使用该类别添加剂不符合技术要求。

5　污染物

本标准所涉及的产品应符合《食品及饲料中污染物和毒素通用标准》（CXS 193-1995）规定的污染物最大限量。

本标准所涉及的产品在加工中使用的牛乳应符合《食品及饲料中污染物和毒素通用标准》（CXS 193-1995）规定的乳中污染物和毒素最大限量以及国际食品法典委员会设定的农药和兽药最大残留限量。

6　卫生要求

建议本标准所涉及的产品应遵循《食品卫生总则》（CXC 1-1969）、《乳及乳制品卫生操作规范》（CXC 57-2004）以及《卫生操作规范》和《生产操作规范》等其他法典文本。产品应符合《食品微生物标准制定与实施原则和准则》（CXG 21-1997）规定的所有微生物标准。

7　标识

除符合《预包装食品标识通用标准》（CXS 1-1985）和《乳品术语》（CXS 206-1999）外，还应符合下列具体规定。

7.1　产品名称

可以根据《预包装食品标识通用标准》（CXS 1-1985）第4.1条使用Edam、Edamer、Edammer等名称。只要产品符合该标准，产品销售国的习惯拼写也可被使用。

只要干酪符合本标准，就可选择使用此名称。对符合本标准但不使用此名称的干酪，应按《干酪通则》（CXS 283-1978）命名规定来命名。

对于脂肪含量超过本标准第3.3条规定的参考含量范围，但高于绝对最低含量的产品，在命名时须附上适当修饰语，作为名称的一部分或处于与名称同一视野内的显眼位置，说明所做调整或脂肪含量（以干物质中脂肪含量或质量百分比表示，视产品零售国的接受情况）。适当修饰语是指《干酪通则》（CXS 283-1978）第7.2条所规定的特征描述用语或是符合《营养和保健声明使用准则》（CXG 23-1997）[1]的营养声明。

该名称也可用于符合本标准的干酪做成的切块、切片、切碎或磨碎产品。

7.2 原产国

应当标记原产国（即生产国，而非名称起源国）。当产品在第二个国家有实质性改造[2]，则以标识目的进行改造的国家应视为原产国。

7.3 乳脂含量说明

乳脂含量应以销售国可接受的方式声明：（i）质量百分比形式；（ii）干物质中脂肪百分比形式；（iii）如果产品标识上标明了份数，也可用每份中乳脂重量（g）表示。

7.4 日期标识

尽管《预包装食品标识通用标准》（CXS 1-1985）第4.7.1条规定，仍应说明生产日期而非最低保存时间信息，只要产品并非直接出售给最终消费者。

7.5 非零售包装标识

在包装容器上除要标明产品名称、批次、生产厂家或包装商的名称和地址外，在包装容器上或附随说明书中，也应按本标准第7条和《预包装食品标识通用标准》（CXS 1-1985）第4.1～4.8条所要求的信息以及储存说明（在必要的地方）加以陈述。如果批次、生产厂家或包装商的名称和地址可以在附随的文件标明，则可以用一个识别标识来代替。

8 抽样和分析方法

为检查产品是否符合本标准，应采用《分析和抽样推荐性方法》（CXS 234-1999）中涉及本标准规定的分析和抽样方法。

---

1 为比较营养声明，干物质中脂肪含量最低参考值为40%。
2 例如，重新包装、切块、切片、切碎和磨碎不构成实质性改变。

附录

附 加 信 息

以下补充信息对前文各条款的规定不构成影响,前文规定内容对产品标识、食品名称的使用以及食品安全性至关重要。

1　外观特征

球状艾丹姆干酪通常重量在1.5~2.5 kg。

2　加工方法

腌制方法:在盐卤水中浸泡。

STANDARD FOR EDAM

CXS 265-1966

Formerly CODEX STAN C-4-1966. Adopted in 1966. Revised in 2007.

Amended in 2008, 2010, 2013, 2018, 2019.

1 SCOPE

This Standard applies to Edam intended for direct consumption or for further processing in conformity with the description in Section 2 of this Standard.

2 DESCRIPTION

Edam is a ripened firm/semi-hard cheese in conformity with the *General Standard for Cheese* (CXS 283-1978). The body has a near white or ivory through to light yellow or yellow colour and a firm-textured (when pressed by thumb) texture, suitable for cutting, with few more or less round rice to pea sized (or mostly up to 10 mm in diameter) gas holes, distributed in a reasonable regular manner throughout the interior of the cheese, but few openings and splits are acceptable. The shape is spherical, of a flat block or of a loaf. The cheese is manufactured and sold with dry rind, which may be coated. Edam of flat block or loaf shape is also sold without rind[1].

For Edam ready for consumption, the ripening procedure to develop flavour and body characteristics is normally from 3 weeks at 10–18 °C depending on the extent of maturity required. Alternative ripening conditions (including the addition of ripening enhancing enzymes) may be used, provided the cheese exhibits similar physical, biochemical and sensory properties as those achieved by the previously stated ripening procedure. Edam intended for further processing need not exhibit the same degree of ripening when justified through technical and/or trade needs.

3 ESSENTIAL COMPOSITION AND QUALITY FACTORS

3.1 Raw materials

Cows' milk or buffaloes' milk, or their mixtures, and products obtained from these milks.

3.2 Permitted ingredients

- Starter cultures of harmless lactic acid and/or flavour producing bacteria and cultures of other harmless micro-organisms;
- Rennet or other safe and suitable coagulating enzymes;
- Sodium chloride and potassium chloride as salt substitute;
- Potable water;
- Safe and suitable enzymes to enhance the ripening process;

---

[1] This is not to mean that the rind has been removed before sale, instead the cheese has been ripened and/or kept in such a way that no rind is developed (a "rindless" cheese). Ripening film is used in the manufacture of rindless cheese. Ripening film may also constitute the coating that protects the cheese. For rindless cheese, see also the Appendix to the *General Standard for Cheese* (CXS 283-1978).

- Safe and suitable processing aids;
- Rice, corn and potato flours and starches: Notwithstanding the provisions in the *General Standard for Cheese* (CXS 283-1978), these substances can be used in the same function as anti-caking agents for treatment of the surface of cut, sliced, and shredded products only, provided they are added only in amounts functionally necessary as governed by Good Manufacturing Practice, taking into account any use of the anti-caking agents listed in Section 4.

3.3 Composition

Milk constituent	Minimum content (m/m)	Maximum content (m/m)	Reference level (m/m)
Milk fat in dry matter	30%	Not restricted	40% to 50%
Dry matter	Depending on the fat in dry matter content, according to the table below		
	Fat in dry matter content (m/m)	**Corresponding minimum dry matter content** (m/m)	
	Equal to or above 30% but less than 40%	47%	
	Equal to or above 40% but less than 45%	51%	
	Equal to or above 45% but less than 50%	55%	
	Equal to or above 50% but less than 60%	57%	
	Equal to or above 60%	62%	

Compositional modifications beyond the minima and maxima specified above for milk fat and dry matter are not considered to be in compliance with section 4.3.3 of the *General Standard for the Use of Dairy Terms* (CXS 206-1999).

4 FOOD ADDITIVES

Only those additives classes indicated as justified in the table below may be used for the product categories specified. Anticaking agents, colours and preservatives used in accordance with Tables 1 and 2 of the *General Standard for Food Additives* (CXS 192-1995) in food category 01.6.2.1 (Ripened cheese, includes rind) and only certain acidity regulators and anticaking agents in Table 3 are acceptable for use in foods conforming to this standard.

Additive functional class	Justified use	
	Cheese mass	Surface/rind treatment
Colours	×[a]	–
Bleaching agents	–	–
Acidity regulators	×	–
Stabilizers	–	–
Thickeners	–	–
Emulsifiers	–	–

Additive functional class	Justified use	
	Cheese mass	Surface/rind treatment
Antioxidants	–	–
Preservatives	×	×
Foaming agents	–	–
Anti–caking agents	–	×(b)

(a) Only to obtain the colour characteristics, as described in Section 2.

(b) For the surface of sliced, cut, shredded or grated cheese, only.

× The use of additives belonging to the class is technologically justified.

– The use of additives belonging to the class is not technologically justified.

5 CONTAMINANTS

The products covered by this Standard shall comply with the Maximum Levels for contaminants that are specified for the product in the *General Standard for Contaminants and Toxins in Food and Feed* (CXS 193-1995).

The milk used in the manufacture of the products covered by this Standard shall comply with the Maximum Levels for contaminants and toxins specified for milk by the *General Standard for Contaminants and Toxins in Food and Feed* (CXS 193-1995) and with the maximum residue limits for veterinary drug residues and pesticides established for milk by the CAC.

6 HYGIENE

It is recommended that the product covered by the provisions of this standard be prepared and handled in accordance with the appropriate sections of the *General Principles of Food Hygiene* (CXC 1-1969), the *Code of Hygienic Practice for Milk and Milk Products* (CXC 57-2004) and other relevant Codex texts such as Codes of Hygienic Practice and Codes of Practice. The products should comply with any microbiological criteria established in accordance with the *Principles and Guidelines for the Establishment and Application of Microbiological Criteria Related to Foods* (CXG 21-1997).

7 LABELLING

In addition to the provisions of the *General Standard for the Labelling of Prepackaged Foods* (CXS 1-1985) and the *General Standard for the Use of Dairy Terms* (CXS 206-1999), the following specific provisions apply.

7.1 Name of the food

The names Edam, Edamer or Edammer may be applied in accordance with Section 4.1 of the *General Standard for the Labelling of Prepackaged Foods* (CXS 1-1985), provided that the product is in conformity with this Standard. Where customary in the country of retail sale, alternative spelling may be used.

The use of the name is an option that may be chosen only if the cheese complies with this standard. Where the name is not used for a cheese that complies with this standard, the naming provisions of the *General Standard for Cheese* (CXS 283-1978) apply.

The designation of products in which the fat content is below or above the reference range but above the absolute minimum specified in Section 3.3 of this Standard shall be accompanied by an appropriate qualification describing the modification made or the fat content (expressed as fat in dry matter or as percentage by mass whichever is acceptable in the country of retail sale), either as part of the name or in a prominent position in the same field of vision. Suitable qualifiers are the appropriate characterizing terms specified in Section 7.2 of the *General Standard for Cheese* (CXS 283-1978) or a nutritional claim in accordance with the *Guidelines for the Use of Nutrition and Health Claims* (CXG 23-1997)[1].

The designation may also be used for cut, sliced, shredded or grated products made from cheese which cheese is in conformity with this Standard.

7.2 Country of origin

The country of origin (which means the country of manufacture, not the country in which the name originated) shall be declared. When the product undergoes substantial transformation[2] in a second country, the country in which the transformation is performed shall be considered to be the country of origin for the purpose of labelling.

7.3 Declaration of milk fat content

The milk fat content shall be declared in a manner found acceptable in the country of retail sale, either (i) as a percentage by mass, (ii) as a percentage of fat in dry matter, or (iii) in grams per serving as quantified in the label, provided that the number of servings is stated.

7.4 Date marking

Notwithstanding the provisions of Section 4.7.1 of the *General Standard for the Labelling of Prepackaged Foods* (CXS 1-1985), the date of manufacture may be declared instead of the minimum durability information, provided that the product is not intended to be purchased as such by the final consumer.

7.5 Labelling of non-retail containers

Information specified in Section 7 of this Standard and Sections 4.1 to 4.8 of the *General Standard for the Labelling of Prepackaged Foods* (CXS 1-1985) and, if necessary, storage instructions, shall be given either on the container or in accompanying documents, except that the name of the product, lot identification, and the name of the manufacturer or packer shall appear on the container, and in the absence of such a container, on the product itself. However, lot identification and the name and address may be replaced by an identification mark, provided that such mark is clearly identifiable with the accompanying documents.

8 METHODS OF SAMPLING AND ANALYSIS

For checking the compliance with this standard, the methods of analysis and sampling contained in the *Recommended Methods of Analysis and Sampling* (CXS 234-1999) relevant to the provisions in this standard, shall be used.

---

1   For the purpose of comparative nutritional claims, the minimum fat content of 40% fat in dry matter constitutes the reference.

2   For instance, repackaging, cutting, slicing, shredding and grating is not regarded as substantial transformation.

APPENDIX

ADDITIONAL INFORMATION

The additional information below does not affect the provisions in the preceding sections which are those that are essential to the product identity, the use of the name of the food and the safety of the food.

1 **Appearance characteristics**

Edam, in the spherical form, is normally manufactured with a weights ranging from 1.5 to 2.5 kg.

2 **Method of manufacture**

Salting method: Salted in brine.

高达干酪

STANDARD FOR GOUDA

CXS 266-1966

原为CODEX STAN C-5-1966。2001年通过。2007年修订。
2008年、2010年、2013年、2018年、2019年修正。

1　范围

本标准适用于符合本标准第2条所述即食或用于进一步加工的高达干酪。

2　说明

高达干酪是指符合《干酪通则》（CXS 283-1978）的成熟硬质或半硬质干酪。干酪外形类似侧面突出的扁平圆柱、平块以及面包状，外观呈白色、银白色、淡黄色、橘黄色，其质地坚硬（用拇指摁压）且适合切割，有少数或多个均匀分布的豆粒、圆针头大小（大多数直接不超过10 mm）气孔，允许个别开口或裂口。干酪加工及销售时带有干皮，也可以有外包装。平块或面包状高达干酪销售时也可以不带皮[1]。对于即食的高达干酪，调制味道和外观特征的催熟程序在10～17 ℃下通常需要3周，取决于所需成熟度。在干酪具备和前述催熟程序同样的物理、生化和感官特性的情况下，可以使用其他催熟条件（包括添加加速催熟的酶）。如有合理技术或贸易需求，用于进一步加工的高达干酪和重量较低的高达干酪（2.5 kg）无需具备同等成熟度。

3　基本成分和质量要求

3.1　基本原料

牛乳、水牛乳或它们的混合乳，以及源于这些乳的制品。

3.2　其他配料

- 无害乳酸发酵剂和/或产香菌和其他无害微生物的培养剂；
- 凝乳酶或其他安全、适宜的凝固酶；
- 作为代盐制品的氯化钠和氯化钾；
- 饮用水；

---

[1] 并非在销售前去掉外皮，相反干酪在成熟后，以不产生外皮方式保存（无皮干酪）。在无皮干酪加工过程中，使用催熟膜。催熟膜也可以成为保护干酪的外衣。无皮干酪也可参见《干酪通则》（CXS 283-1978）附件。

- 促进成熟的安全、适宜的酶；
- 安全、适宜的加工助剂；
- 大米、玉米、马铃薯粉和淀粉：尽管有《干酪通则》（CXS 283-1978），但这些物质同样可作为抗结块剂处理切块、切片和切碎产品表面，只要添加量是良好操作规范所要求达到期待功能的必需量，同时考虑第4条规定的抗结块剂的使用。

3.3　成分

乳成分	最低含量（m/m）	最高含量（m/m）	参考含量（m/m）
干物质乳脂	30%	/	48%~55%
干物质	根据干物质脂肪含量，参见下表		
	干物质脂肪含量（m/m）	相应最小干物质含量（m/m）	
	≥30%，<40%	48%	
	≥40%，<48%	52%	
	≥48%，<60%	55%	
	≥60%	62%	

FDM（干物质脂肪含量）在40%~48%和重量小于2.5 kg的高达干酪可以在DM（干物质）含量不超过50%的情况下销售，前提是名称符合术语"小"。

乳脂和干物质含量超过规定的最大值和最小值被视为不符合《乳品术语》（CXS 206-1999）第4.3.3条。

4　食品添加剂

只有下表中标注为合理使用的添加剂类别才可用于指定产品类别。根据《食品添加剂通用标准》（CXS 192-1995）表1和表2在食品类别01.6.2.1（成熟干酪，包括外皮）中使用的抗结块剂、着色剂和防腐剂，以及表3中仅部分酸度调节剂和抗结块剂可用于符合本标准的食品。

添加剂功能类别	合理使用	
	干酪量	表面/外皮处理
着色剂	×[a]	–
漂白剂	–	–
酸度调节剂	×	–
稳定剂	–	–

添加剂功能类别	合理使用	
	干酪量	表面/外皮处理
增稠剂	-	-
乳化剂	-	-
抗氧化剂	-	-
防腐剂	×	×
发泡剂	-	-
抗结块剂	-	×[b]

（a）仅为获得第2条所述颜色特征。
（b）仅用于切片、切块、切碎或磨碎的乳酪表面。
× 使用该类别添加剂符合技术要求。
- 使用该类别添加剂不符合技术要求。

5 污染物

本标准所涉及的产品应符合《食品及饲料中污染物和毒素通用标准》（CXS 193-1995）规定的污染物最大限量。

本标准所涉及的产品在加工中使用的牛乳应符合《食品及饲料中污染物和毒素通用标准》（CXS 193-1995）规定的乳中污染物和毒素最大限量以及国际食品法典委员会设定的农药和兽药最大残留限量。

6 卫生要求

建议本标准所涉及的产品应遵循《食品卫生总则》（CXC 1-1969）、《乳及乳制品卫生操作规范》（CXC 57-2004）以及《卫生操作规范》和《生产操作规范》等其他相关法典文本。产品应符合《食品微生物标准制定与实施原则和准则》（CXG 21-1997）规定的所有微生物标准。

7 标识

除符合《预包装食品标识通用标准》（CXS 1-1985）和《乳品术语》（CXS 206-1999）外，还应符合下列具体规定。

7.1 产品名称

高达干酪名称应符合《预包装食品标识通用标准》（CXS 1-1985）第4.1条，只要产品符合该标准，产品销售国的习惯拼写也可被使用。

对于脂肪含量超过本标准第3.3条规定的参考含量范围，但高于绝对最低含量的产品，在命名时须附上适当修饰语，作为名称的一部分或处于与名称同一视野内的显眼位置，说明所做调整或脂肪含量（以干物质中脂肪含量或质量百分比表示，视产品销售国的接受情况）。适当修饰语是指《干酪通则》（CXS 283-1978）第7.2条所规定的特征描述用语或是符合《营养和保健声明使用准则》（CXG 23-1997）[1]的营养声明。

该名称也可用于符合本标准的干酪做成的切块、切片、切碎或磨碎产品。

7.2 原产国

应当标记原产国（即生产国，而非名称起源国）。当产品在第二个国家有实质性改造[2]，则以标识目的进行改造的国家应视为原产国。

7.3 乳脂含量说明

乳脂含量应以销售国可接受的方式声明：（i）质量百分比形式；（ii）干物质中脂肪百分比形式；（iii）如果产品标识上标明了份数，也可用每份中乳脂重量（g）表示。

7.4 日期标识

尽管《预包装食品标识通用标准》（CXS 1-1985）第4.7.1条规定，仍应说明生产日期而非最低保存时间信息，只要产品并非直接出售给最终消费者。

7.5 非零售包装标识

在包装容器上除要标明产品名称、批次、生产厂家或包装商的名称和地址外，在包装容器上或附随说明书中，也应按本标准第7条和《预包装食品标识通用标准》（CXS 1-1985）第4.1~4.8条所要求的信息以及储存说明（在必要的地方）加以陈述。如果批次、生产厂家或包装商的名称和地址可以在附随的文件标明，则可以用一个识别标识来代替。

8 抽样和分析方法

为检查产品是否符合本标准，应采用《分析和抽样推荐性方法》（CXS 234-1999）中涉及本标准规定的分析和抽样方法。

---

[1] 为比较营养声明，干物质中脂肪含量最低参考值为48%。
[2] 例如，重新包装、切块、切片、切碎和磨碎不构成实质性改变。

附录

附 加 信 息

以下补充信息对前文各条款的规定不构成影响,前文规定内容对产品标识、食品名称的使用以及食品安全性至关重要。

1 外观特征

高达干酪通常重量范围是2.5~30 kg,更低重量通常被称为术语"小"。

2 加工方法

腌制方法:在盐卤水中浸泡。

STANDARD FOR GOUDA

CXS 266-1966

Formerly CODEX STAN C-5-1966. Adopted in 2001. Revised in 2007.

Amended in 2008, 2010, 2013, 2018, 2019.

1 SCOPE

This Standard applies to Gouda intended for direct consumption or for further processing in conformity with the description in Section 2 of this Standard.

2 DESCRIPTION

Gouda is a ripened firm/semi-hard cheese in conformity with the *General Standard for Cheese* (CXS 283-1978). The body has a near white or ivory through to light yellow or yellow colour and a firm-textured (when pressed by thumb) texture, suitable for cutting, with few to plentiful, more or less round pin's head to pea sized (or mostly up to 10 mm in diameter) gas holes, distributed in a reasonable regular manner throughout the interior of the cheese, but few openings and splits are acceptable. The shape is of a flattened cylinder with convex sides, a flat block, or a loaf. The cheese is manufactured and sold with a dry rind, which may be coated. Gouda of flat block or loaf shape is also sold without[1] rind.

For Gouda ready for consumption, the ripening procedure to develop flavour and body characteristics is normally from 3 weeks at 10–17 °C depending on the extent of maturity required. Alternative ripening conditions (including the addition of ripening enhancing enzymes) may be used, provided the cheese exhibits similar physical, biochemical and sensory properties as those achieved by the previously stated ripening procedure. Gouda intended for further processing and Gouda of low weights (< 2.5 kg) need not exhibit the same degree of ripening when justified through technical and/or trade needs.

3 ESSENTIAL COMPOSITION AND QUALITY FACTORS

3.1 Raw materials

Cows' milk or buffaloes' milk, or their mixtures, and products obtained from these milks.

3.2 Permitted ingredients

– Starter cultures of harmless lactic acid and/or flavour producing bacteria and cultures of other harmless micro-organisms;
– Rennet or other safe and suitable coagulating enzymes;
– Sodium chloride; and potassium chloride as a salt substitute;
– Potable water;
– Safe and suitable enzymes to enhance the ripening process;

---

1 This is not to mean that the rind has been removed before sale, instead the cheese has been ripened and/or kept in such a way that no rind is developed (a "rindless" cheese). Ripening film is used in the manufacture of rindless cheese. Ripening film may also constitute the coating that protects the cheese. For rindless cheese, see also the Appendix to the *General Standard for Cheese* (CXS 283-1978).

- Safe and suitable processing aids;
- Rice, corn and potato flours and starches: Notwithstanding the provisions in the *General Standard for Cheese* (CXS 283-1978), these substances can be used in the same function as anti-caking agents for treatment of the surface of cut, sliced, and shredded products only, provided they are added only in amounts functionally necessary as governed by Good Manufacturing Practice, taking into account any use of the anti-caking agents listed in Section 4.

3.3 Composition

Milk constituent	Minimum content (m/m)	Maximum content (m/m)	Reference level (m/m)
Milk fat in dry matter	30%	Not restricted	48% to 55%
Dry matter	Depending on the fat in dry matter content, according to the table below		
	Fat in dry matter content (m/m)	**Corresponding minimum dry matter content** (m/m)	
	Equal to or above 30% but less than 40%	48%	
	Equal to or above 40% but less than 48%	52%	
	Equal to or above 48% but less than 60%	55%	
	Equal to or above 60%	62%	

Gouda with between 40% and 48% FDM and with a weight of less than 2.5 kg can be sold with a DM content of min. 50%, provided that the name is qualified by the term "baby".

Compositional modifications beyond the minima and maxima specified above for milk fat and dry matter are not considered to be in compliance with section 4.3.3 of the *General Standard for the Use of Dairy Terms* (CXS 206-1999).

4 FOOD ADDITIVES

Only those additives classes indicated as justified in the table below may be used for the product categories specified. Anticaking agents, colours and preservatives used in accordance with Tables 1 and 2 of the *General Standard for Food Additives* (CXS 192-1995) in food category 01.6.2.1 (Ripened cheese, includes rind) and only certain acidity regulators and anticaking agents in Table 3 are acceptable for use in foods conforming to this standard.

Additive functional class	Justified use	
	Cheese mass	Surface/rind treatment
Colours	x[a]	–
Bleaching agents	–	–
Acidity regulators	x	–
Stabilizers	–	–
Thickeners	–	–
Emulsifiers	–	–

Additive functional class	Justified use	
	Cheese mass	Surface/rind treatment
Antioxidants	–	–
Preservatives	×	×
Foaming agents	–	–
Anti–caking agents	–	×(b)

(a) Only to obtain the colour characteristics, as described in Section 2.

(b) For the surface of sliced, cut, shredded or grated cheese, only.

× The use of additives belonging to the class is technologically justified.

– The use of additives belonging to the class is not technologically justified.

5 CONTAMINANTS

The products covered by this Standard shall comply with the Maximum Levels for contaminants that are specified for the product in the *General Standard for Contaminants and Toxins in Food and Feed* (CXS 193-1995).

The milk used in the manufacture of the products covered by this Standard shall comply with the Maximum Levels for contaminants and toxins specified for milk by the *General Standard for Contaminants and Toxins in Food and Feed* (CXS 193-1995) and with the maximum residue limits for veterinary drug residues and pesticides established for milk by the CAC.

6 HYGIENE

It is recommended that the product covered by the provisions of this standard be prepared and handled in accordance with the appropriate sections of the *General Principles of Food Hygiene* (CXC 1-1969), the *Code of Hygienic Practice for Milk and Milk Products* (CXC 57-2004) and other relevant Codex texts such as Codes of Hygienic Practice and Codes of Practice. The products should comply with any microbiological criteria established in accordance with the *Principles and Guidelines for the Establishment and Application of Microbiological Criteria Related to Foods* (CXG 21-1997).

7 LABELLING

In addition to the provisions of the *General Standard for the Labelling of Prepackaged Foods* (CXS 1-1985) and the *General Standard for the Use of Dairy Terms* (CXS 206-1999), the following specific provisions apply.

7.1 Name of the food

The name Gouda may be applied in accordance with Section 4.1 of the *General Standard for the Labelling of Prepackaged Foods* (CXS 1-1985), provided that the product is in conformity with this Standard. Where customary in the country of retail sale, alternative spelling may be used.

The use of the name is an option that may be chosen only if the cheese complies with this standard. Where the name is not used for a cheese that complies with this standard, the naming provisions of the *General Standard for Cheese* (CXS 283-1978) apply.

The designation of products in which the fat content is below or above the reference

range but above the absolute minimum specified in section 3.3 of this Standard shall be accompanied by an appropriate qualification describing the modification made or the fat content (expressed as fat in dry matter or as percentage by mass whichever is acceptable in the country of retail sale), either as part of the name or in a prominent position in the same field of vision. Suitable qualifiers are the appropriate characterizing terms specified in Section 7.2 of the *General Standard for Cheese* (CXS 283-1978) or a nutritional claim in accordance with the *Guidelines for Use of Nutrition and Health Claims* (CXG 23-1997)[1].

The designation may also be used for cut, sliced, shredded or grated products made from cheese which cheese is in conformity with this Standard.

7.2 Country of origin

The country of origin (which means the country of manufacture, not the country in which the name originated) shall be declared. When the product undergoes substantial transformation[2] in a second country, the country in which the transformation is performed shall be considered to be the country of origin for the purpose of labelling.

7.3 Declaration of milk fat content

The milk fat content shall be declared in a manner found acceptable in the country of retail sale, either (i) as a percentage by mass, (ii) as a percentage of fat in dry matter, or (iii) in grams per serving as quantified in the label, provided that the number of servings is stated.

7.4 Date marking

Notwithstanding the provisions of Section 4.7.1 of the *General Standard for the Labelling of Prepackaged Foods* (CXS 1-1985), the date of manufacture may be declared instead of the minimum durability information, provided that the product is not intended to be purchased as such by the final consumer.

7.5 Labelling of non-retail containers

Information specified in Section 7 of this Standard and Sections 4.1 to 4.8 of the *General Standard for the Labelling of Prepackaged Foods* (CXS 1-1985) and, if necessary, storage instructions, shall be given either on the container or in accompanying documents, except that the name of the product, lot identification, and the name of the manufacturer or packer shall appear on the container, and in the absence of such a container, on the product itself. However, lot identification and the name and address may be replaced by an identification mark, provided that such mark is clearly identifiable with the accompanying documents.

8 METHODS OF SAMPLING AND ANALYSIS

For checking the compliance with this standard, the methods of analysis and sampling contained in the *Recommended Methods of Analysis and Sampling* (CXS 234-1999) relevant to the provisions in this standard, shall be used.

---

[1] For the purpose of comparative nutritional claims, the minimum fat content of 48% fat in dry matter constitutes the reference.

[2] For instance, repackaging, cutting, slicing, shredding and grating is not regarded as substantial transformation.

APPENDIX

ADDITIONAL INFORMATION

The additional information below does not affect the provisions in the preceding sections which are those that are essential to the product identity, the use of the name of the food and the safety of the food.

1 **Appearance characteristics**

Gouda is normally manufactured with weights ranging from 2.5 to 30 kg. Lower weights are normally qualified by the term "Baby".

2 **Method of manufacture**

Salting method: Salted in brine.

哈瓦蒂干酪

STANDARD FOR HAVARTI

CXS 267-1966

原为CODEX STAN C-6-1966。1966年通过。2007年修订。
2008年、2010年、2013年、2018年、2019年修正。

1 范围

本标准适用于符合本标准第2条所述即食或用于进一步加工的哈瓦蒂干酪。

2 说明

哈瓦蒂干酪是指符合《干酪通则》（CXC 283-1978）的成熟硬质或半硬质干酪发酵。干酪外形呈平柱状、扁平状或矩形，外观呈白色、银白色、淡黄色或橘黄色，其质地适宜切割，有大量不规则米粒大小（宽1~2 mm，长度最多10 mm）气孔。干酪在销售时，轻微油性的成熟外皮将干酪包裹起来，不带外皮也可。但部分商品干酪也不具有这样的表皮[1]。

对于即食的哈瓦蒂干酪，发酵时间根据所需成熟程度而定，在14~18 ℃下发酵1~2周（产生斑点），紧接着在8~12 ℃下发酵1~3周，成熟度取决于因受发酵时间影响的重量。可以使用替代的催熟条件（包括添加催熟酶），前提是奶酪表现出与之前所述的成熟过程所达到的物理、生化和感官特性相似的特性。对于需进一步深加工的哈瓦蒂干酪，如有合理技术或贸易需求，无需上述催熟程序。

3 基本成分和质量要求

3.1 基本原料

牛乳、水牛乳或它们的混合乳，以及源于这些乳的制品。

3.2 其他配料

- 无害乳酸发酵剂和/或产香菌和其他无害微生物的培养剂；
- 凝乳酶或其他安全、适宜的凝固酶；
- 作为代替制品的氯化钠和氯化钾；
- 饮用水；

---

1 并非在销售前去掉外皮，相反干酪在成熟后，以不产生外皮方式保存（无皮干酪）。在无皮干酪加工过程中，使用催熟膜。催熟膜也可以成为保护干酪的外衣。无皮干酪也可参见《干酪通则》附件（CXS 283-1978）。

- 促进成熟的安全、适宜的酶；
- 安全、适宜的加工助剂；
- 大米、玉米、马铃薯粉和淀粉：尽管有《干酪通则》（CXS 283-1978），但这些物质同样可作为抗结块剂处理切块、切片和切碎产品表面，只要添加量是良好操作规范所要求达到期待功能的必需量，同时考虑第4条规定的抗结块剂使用。

3.3 成分

乳成分	最低含量（m/m）	最高含量（m/m）	参考含量（m/m）
干物质乳脂	30%	/	45%~55%
干物质	根据干物质脂肪含量，参见下表		
	干物质脂肪含量（m/m）	相应最低干物质含量（m/m）	
	≥30%，<40%	46%	
	≥40%，<45%	48%	
	≥45%，<55%	50%	
	≥55%，<60%	54%	
	≥60%	58%	

乳脂和干物质含量超过规定的最大值和最小值被视为不符合《乳品术语》（CXS 206-1999）第4.3.3条。

4 食品添加剂

只有下表中列出的食品添加剂种类允许在干酪生产中使用。根据《食品添加剂通用标准》（CXS 192-1995）表1和表2在食品类别01.6.2.1（成熟干酪，包括外皮）中使用的抗结块剂、着色剂和防腐剂，以及表3中仅部分酸度调节剂和抗结块剂可用于符合本标准的食品。

添加剂功能类别	合理使用	
	干酪量	表面/外皮处理
着色剂	×[a]	—
漂白剂	—	—
酸度调节剂	×	—
稳定剂	—	—

添加剂功能类别	合理使用	
	干酪量	表面/外皮处理
增稠剂	-	-
乳化剂	-	-
抗氧化剂	-	-
防腐剂	×	×
发泡剂	-	-
抗结块剂	-	×[b]

（a）仅为获得第2条所述颜色特征。
（b）仅用于切片、切块、切碎或磨碎的乳酪表面。
× 使用该类别添加剂符合技术要求。
- 使用该类别添加剂不符合技术要求。

5　污染物

本标准所涉及的产品应符合《食品及饲料中污染物和毒素通用标准》（CXS 193-1995）规定的污染物最大限量。

本标准所涉及的产品在加工中使用的牛乳应符合《食品及饲料中污染物和毒素通用标准》（CXS 193-1995）规定的乳中污染物和毒素最大限量以及国际食品法典委员会设定的农药和兽药最大残留限量。

6　卫生要求

建议本标准所涉及的产品应遵循《食品卫生总则》（CXC 1-1969）、《乳及乳制品卫生操作规范》（CXC 57-2004）以及《卫生操作规范》和《生产操作规范》等其他相关法典文本。产品应符合《食品微生物标准制定与实施原则和准则》（CXG 21-1997）规定的所有微生物标准。

7　标识

除符合《预包装食品标识通用标准》（CXS 1-1985）和《乳品术语》（CXS 206-1999）外，还应符合下列具体规定。

7.1　产品名称

哈瓦蒂干酪产品名称应符合《预包装食品标识通用标准》（CXS 1-1985）第4.1条，只要产品符合该标准，产品销售国的习惯拼写也可被使用。

只要干酪符合本标准，就可选择使用此名称。对符合本标准但不使用此名称的干酪，应按《干酪通则》（CXS 283-1978）命名规则来命名。

对于脂肪含量超过本标准第3.3条规定的参考含量范围，但高于绝对最低含量的产品，在命名时须附上适当修饰语，作为名称的一部分或处于与名称同一视野内的显眼位置，说明所做调整或脂肪含量（以干物质中脂肪含量或质量百分比表示，视产品销售国的接受情况）。适当修饰语是指《干酪通则》（CXS 283-1978）第7.2条所规定的特征描述用语或是符合《营养和保健声明使用准则》（CXG 23-1997）[1]的营养声明。

该名称也可用于符合本标准规定的干酪做成的切块、切片、切碎或磨碎产品。

7.2 原产国

应当标记原产国（即生产国，而非名称起源国）。当产品在第二个国家有实质性改造[2]，则以标识目的进行改造的国家应视为原产国。

7.3 乳脂含量说明

乳脂含量应以销售国可接受的方式声明：（i）质量百分比形式；（ii）干物质中脂肪百分比形式；（iii）如果产品标识上标明了份数，也可用每份中乳脂重量（g）表示。

7.4 日期标识

尽管《预包装食品标识通用标准》（CXS 1-1985）第4.7.1条规定，仍应说明生产日期而非最低保存时间信息，只要产品并非直接出售给最终消费者。

7.5 非零售包装标识

在包装容器上除要标明产品名称、批次、生产厂家或包装商的名称和地址外，在包装容器上或附随说明书中，也应按本标准第7条和《预包装食品标识通用标准》（CXS 1-1985）第4.1~4.8条所要求的信息以及储存说明（在必要的地方）加以陈述。如果批次、生产厂家或包装商的名称和地址可以在附随的文件标明，则可以用一个识别标识来代替。

8 抽样和分析方法

为检查产品是否符合本标准，应采用《分析和抽样推荐性方法》（CXS 234-1999）中涉及本标准规定的分析和抽样方法。

---

1 为比较营养声明，干物质中脂肪含量最低参考值为45%。
2 例如，重新包装、切块、切片、切碎和磨碎不构成实质性改变。

STANDARD FOR HAVARTI

CXS 267-1966

Formerly CODEX STAN C-6-1966. Adopted in 1966. Revised in 2007.

Amended in 2008, 2010, 2013, 2018, 2019.

1 SCOPE

This Standard applies to Havarti intended for direct consumption or for further processing in conformity with the description in Section 2 of this Standard.

2 DESCRIPTION

Havarti is a ripened firm/semi-hard cheese in conformity with the *General Standard for Cheese* (CXS 283-1978). The body has a near white or ivory through to light yellow or yellow colour and a texture suitable for cutting, with plentiful, irregular and coarse large rice seed sized (or mostly 1–2 mm in width and up to 10 mm in length) gas holes. The shape is flat cylindrical, rectangular or of a loaf shape. The cheese is sold with or without[1] a slightly greasy smear ripened rind, which may be coated.

For Havarti ready for consumption, the ripening procedure to develop flavour and body characteristics is normally, depending on weight, 1–2 weeks at 14–18 °C (for smear development) followed by from 1–3 weeks at 8–12 °C depending on the extent of maturity required. Alternative ripening conditions (including the addition of ripening enhancing enzymes) may be used, provided the cheese exhibits similar physical, biochemical and sensory properties as those achieved by the previously stated ripening procedure. Havarti intended for further processing need not exhibit the same degree of ripening when justified through technical and/or trade needs.

3 ESSENTIAL COMPOSITION AND QUALITY FACTORS

3.1 Raw materials

Cows' milk or buffaloes' milk, or their mixtures, and products obtained from these milks.

3.2 Permitted ingredients

- Starter cultures of harmless lactic acid and/or flavour producing bacteria and cultures of other harmless micro-organisms;
- Rennet or other safe and suitable coagulating enzymes;
- Sodium chloride and potassium chloride as a salt substitute;
- Potable water;
- Safe and suitable enzymes to enhance the ripening process;

---

[1] This is not to mean that the rind has been removed before sale, instead the cheese has been ripened and/or kept in such a way that no rind is developed (a "rindless" cheese). Ripening film is used in the manufacture of rindless cheese. Ripening film may also constitute the coating that protects the cheese. For rindless cheese see also the Appendix to the *General Standard for Cheese* (CXS 283-1978).

- Safe and suitable processing aids;
- Rice, corn and potato flours and starches: Notwithstanding the provisions in the *General Standard for Cheese* (CXS 283-1978), these substances can be used in the same function as anti-caking agents for treatment of the surface of cut, sliced, and shredded products only, provided they are added only in amounts functionally necessary as governed by Good Manufacturing Practice, taking into account any use of the anti-caking agents listed in Section 4.

3.3 Composition

Milk constituent	Minimum content (m/m)	Maximum content (m/m)	Reference level (m/m)
Milk fat in dry matter	30%	Not restricted	45% to 55%
Dry matter	Depending on the fat in dry matter content, according to the table below		
	Fat in dry matter content (m/m)	Corresponding minimum dry matter content (m/m)	
	Equal to or above 30% but less than 40%	46%	
	Equal to or above 40% but less than 45%	48%	
	Equal to or above 45% but less than 55%	50%	
	Equal to or above 55% but less than 60%	54%	
	Equal to or above 60%	58%	

Compositional modifications beyond the minima and maxima specified above for milk fat and dry matter are not considered to be in compliance with section 4.3.3 of the *General Standard for the Use of Dairy Terms* (CXS 206-1999).

4 FOOD ADDITIVES

Only those additives classes indicated as justified in the table below may be used for the product categories specified. Anticaking agents, colours and preservatives used in accordance with Tables 1 and 2 of the *General Standard for Food Additives* (CXS 192-1995) in food category 01.6.2.1 (Ripened cheese, includes rind) and only certain acidity regulators and anticaking agents in Table 3 are acceptable for use in foods conforming to this standard.

Additive functional class	Justified use	
	Cheese mass	Surface/rind treatment
Colours	×[a]	–
Bleaching agents	–	–
Acidity regulators	×	–
Stabilizers	–	–
Thickeners	–	–
Emulsifiers	–	–

Additive functional class	Justified use	
	Cheese mass	Surface/rind treatment
Antioxidants	–	–
Preservatives	×	×
Foaming agents	–	–
Anti–caking agents	–	×(b)

(a) Only to obtain the colour characteristics, as described in Section 2.
(b) For the surface of sliced, cut, shredded or grated cheese, only.
× The use of additives belonging to the class is technologically justified.
– The use of additives belonging to the class is not technologically justified.

5 CONTAMINANTS

The products covered by this Standard shall comply with the Maximum Levels for contaminants that are specified for the product in the *General Standard for Contaminants and Toxins in Food and Feed* (CXS 193-1995).

The milk used in the manufacture of the products covered by this Standard shall comply with the Maximum Levels for contaminants and toxins specified for milk by the *General Standard for Contaminants and Toxins in Food and Feed* (CXS 193-1995) and with the maximum residue limits for veterinary drug residues and pesticides established for milk by the CAC.

6 HYGIENE

It is recommended that the product covered by the provisions of this standard be prepared and handled in accordance with the appropriate sections of the *General Principles of Food Hygiene* (CXC 1-1969), the *Code of Hygienic Practice for Milk and Milk Products* (CXC 57-2004) and other relevant Codex texts such as Codes of Hygienic Practice and Codes of Practice. The products should comply with any microbiological criteria established in accordance with the *Principles and Guidelines for the Establishment and Application of Microbiological Criteria Related to Foods* (CXG 21-1997).

7 LABELLING

In addition to the provisions of the *General Standard for the Labelling of Prepackaged Foods* (CXS 1-1985) and the *General Standard for the Use of Dairy Terms* (CXS 206-1999), the following specific provisions apply.

7.1 Name of the food

The name Havarti may be applied in accordance with Section 4.1 of the *General Standard for the Labelling of Prepackaged Foods* (CXS 1-1985), provided that the product is in conformity with this Standard. Where customary in the country of retail sale, alternative spelling may be used.

The use of the name is an option that may be chosen only if the cheese complies with this standard. Where the name is not used for a cheese that complies with this standard, the naming provisions of the *General Standard for Cheese* (CXS 283-1978) apply.

The designation of products in which the fat content is below or above the reference

range but above the absolute minimum specified in section 3.3 of this Standard shall be accompanied by an appropriate qualification describing the modification made or the fat content (expressed as fat in dry matter or as percentage by mass whichever is acceptable in the country of retail sale), either as part of the name or in a prominent position in the same field of vision. Suitable qualifiers are the appropriate characterizing terms specified in Section 7.2 of the *General Standard for Cheese* (CXS 283-1978) or a nutritional claim in accordance with the *Guidelines for Use of Nutrition and Health Claims* (CXG 23-1997)[1].

The designation may also be used for cut, sliced, shredded or grated products made from cheese which cheese is in conformity with this Standard.

7.2 Country of origin

The country of origin (which means the country of manufacture, not the country in which the name originated) shall be declared. When the product undergoes substantial transformation[2] in a second country, the country in which the transformation is performed shall be considered to be the country of origin for the purpose of labelling.

7.3 Declaration of milk fat content

The milk fat content shall be declared in a manner found acceptable in the country of retail sale either (i) as a percentage by mass, (ii) as a percentage of fat in dry matter, or (iii) in grams per serving as quantified in the label, provided that the number of servings is stated.

7.4 Date marking

Notwithstanding the provisions of Section 4.7.1 of the *General Standard for the Labelling of Prepackaged Foods* (CXS 1-1985), the date of manufacture may be declared instead of the minimum durability information, provided that the product is not intended to be purchased as such by the final consumer.

7.5 Labelling of non-retail containers

Information specified in Section 7 of this Standard and Sections 4.1 to 4.8 of the *General Standard for the Labelling of Prepackaged Foods* (CXS 1-1985) and, if necessary, storage instructions, shall be given either on the container or in accompanying documents, except that the name of the product, lot identification, and the name of the manufacturer or packer shall appear on the container, and in the absence of such a container, on the product itself. However, lot identification and the name and address may be replaced by an identification mark, provided that such mark is clearly identifiable with the accompanying documents.

8 METHODS OF SAMPLING AND ANALYSIS

For checking the compliance with this standard, the methods of analysis and sampling contained in the *Recommended Methods of Analysis and Sampling* (CXS 234-1999) relevant to the provisions in this standard, shall be used.

---

1 For the purpose of comparative nutritional claims, the minimum fat content of 45% fat in dry matter constitutes the reference.

2 For instance, repackaging, cutting, slicing, shredding and grating is not regarded as substantial transformation.

萨姆索干酪

STANDARD FOR SAMSØ

CXS 268-1966

原为CODEX STAN C-7-1966，1966年通过，2007年修订，
2008年、2010年、2013年、2018年、2019年修正

1 范围

本标准适用于符合本标准第2条所述即食或用于进一步加工的萨姆索干酪。

2 说明

萨姆索干酪是指符合《干酪通则》（CXS 283-1978）的成熟硬质干酪。干酪外形呈平柱状、扁平状或矩形，外观呈白色、银白色、淡黄色、橘黄色，其质地坚硬（用拇指摁压）、适合切割，有少数或多个均匀分布、光滑、似圆豆及樱桃大小（大多数直径不超过20 mm）的气孔，允许个别开口或裂口。干酪在销售时干燥、坚硬的外皮将干酪包裹起来，不带外皮也可[1]。

对于即食的萨姆索干酪，调制味道和外观特征的催熟程序在8~17 ℃下通常需要3周，取决于所需成熟度。在干酪具备和前述催熟程序同样的物理、生化和感官特性情况下可以使用其他催熟条件（包括添加加速催熟的酶）。如有合理技术或贸易需求，用于进一步加工的萨姆索干酪无需具备同等成熟度。

3 基本成分和质量要求

3.1 基本原料

牛乳、水牛乳或它们的混合乳，以及源于这些乳的制品。

3.2 其他配料

- 无害乳酸发酵剂和/或产香菌和其他无害微生物的培养剂；
- 凝乳酶或其他安全、适宜的凝固酶；
- 作为代盐制品的氯化钠和氯化钾；
- 饮用水；
- 促进成熟的安全、适宜的酶；
- 安全、适宜的加工助剂；

---

[1] 并非在销售前去掉外皮，相反干酪在成熟后，以不产生外皮方式保存（无皮干酪）。在无皮干酪加工过程中，使用催熟膜。催熟膜也可以成为保护干酪的外衣。无皮干酪也可参见《干酪通则》（CXS 283-1978）附录。

— 大米、玉米、马铃薯粉和淀粉：尽管有《干酪通则》（CXS 283-1978），但这些物质同样可作为抗结块剂处理切块、切片和切碎产品表面，只要添加量是良好操作规范所要求达到期待功能的必需量，同时考虑第4条规定的抗结块剂使用。

3.3 成分

乳成分	最低含量（m/m）	最高含量（m/m）	参考含量（m/m）
干物质乳脂	30%	/	45%~55%
干物质	根据干物质脂肪含量，参见下表		
	干物质脂肪含量（m/m）	相应最低干物质含量（m/m）	
	≥30%，<40%	46%	
	≥40%，<45%	52%	
	≥45%，<55%	54%	
	≥55%	59%	

乳脂和干物质含量超过规定的最大值和最小值被视为不符合《乳品术语》（CXS 206-1999）第4.3.3条。

4 食品添加剂

只有下表中标注为合理使用的添加剂类别才可以用于指定产品类别。根据《食品添加剂通用标准》（CXS 192-1995）表1和表2在食品类别01.6.2.1（成熟干酪，包括外皮）中使用的抗结块剂、着色剂和防腐剂，以及表3中仅部分酸度调节剂和抗结块剂可用于符合本标准的食品。

添加剂功能类别	合理使用	
	干酪量	表面/外皮处理
着色剂	×(a)	—
漂白剂	—	—
酸度调节剂	×	—
稳定剂	—	—
增稠剂	—	—
乳化剂	—	—
抗氧化剂	—	—
防腐剂	×	×
发泡剂	—	—

添加剂功能类别	合理使用	
	干酪量	表面/外皮处理
抗结块剂	−	×(b)

（a）仅为获得第2条所述颜色特征。
（b）仅用于切片、切块、切碎或磨碎的乳酪表面。
× 使用该类别添加剂符合技术要求。
− 使用该类别添加剂不符合技术要求。

5 污染物

本标准所涉及的产品应符合《食品及饲料中污染物和毒素通用标准》（CXS 193-1995）规定的污染物最大限量。

本标准所涉及的产品在加工中使用的牛乳应符合《食品及饲料中污染物和毒素通用标准》（CXS 193-1995）规定的乳中污染物和毒素最大限量以及国际食品法典委员会设定的农药和兽药最大残留限量。

6 卫生要求

建议本标准所涉及的产品应遵循《食品卫生总则》（CXC 1-1969）、《乳及乳制品卫生操作规范》（CXC 57-2004）以及《卫生操作规范》和《生产操作规范》等其他相关法典文本。产品应符合《食品微生物标准制定与实施原则和准则》（CXG 21-1997）规定的所有微生物标准。

7 标识

除符合《预包装食品标识通用标准》（CXS 1-1985）和《乳品术语》（CXS 206-1999）外，还应符合下列具体规定。

7.1 产品名称

萨姆索干酪应符合《预包装食品标识通用标准》（CXS 1-1985）第4.1条，只要产品符合该标准，产品销售国的习惯拼写也可被使用。

只要干酪符合本标准，就可选择使用此名称。对符合本标准但不使用此名称的干酪，应按《干酪通则》（CXS 283-1978）命名规定来命名。

对于脂肪含量超过本标准第3.3条规定的参考含量范围，但高于绝对最低含量的产品，在命名时须附上适当修饰语，作为名称的一部分或处于与名称同一视野内的显眼位置，说明所做调整或脂肪含量（以干物质中脂肪含量或质量百分比表示，视产品销售国的接受情况）。适当修饰语是指《干酪通则》（CXS 283-1978）第7.2条所规定的特征描述用语或是符合《营养和保健声明使用准则》（CXG 23-1997）[1]的

---

[1] 为比较营养声明，干物质中脂肪含量最低参考值为45%。

营养声明。

该名称也可用于符合本标准的干酪做成的切块、切片、切碎或磨碎产品。

7.2 原产国

应当标记原产国（即生产国，而非名称起源国）。当产品在第二个国家有实质性改造[1]，则以标识为目的进行改造的国家应视为原产国。

7.3 乳脂含量说明

乳脂含量应以销售国可以接受的方式声明：（i）质量百分比形式；（ii）干物质中脂肪百分比形式；（iii）如果产品标识上标明了份数，也可用每份中乳脂重量（g）表示。

7.4 日期标识

尽管《预包装食品标识通用标准》（CXS 1-1985）第4.7.1条规定，仍应说明生产日期而非最低保存时间信息，只要产品并非直接出售给最终消费者。

7.5 非零售包装标识

在包装容器上除要标明产品名称、批次、生产厂家或包装商的名称和地址外，在包装容器上或附随说明书中，也应按本标准第7条和《预包装食品标识通用标准》（CXS 1-1985）第4.1～4.8条所要求的信息以及储存说明（在必要的地方）加以陈述。如果批次、生产厂家或包装商的名称和地址可以在附随的文件标明，则可以用一个识别标识来代替。

8 抽样和分析方法

为检查产品是否符合本标准，应采用《分析和抽样推荐性方法》（CXS 234-1999）中涉及本标准规定的分析和抽样方法。

1　例如，重新包装、切块、切片、切碎和磨碎不构成实质性改变。

STANDARD FOR SAMSØ

CXS 268-1966

Formerly CODEX STAN C-7-1966. Adopted in 1966. Revised in 2007.

Amended in 2008, 2010, 2013, 2018, 2019.

1 SCOPE

This Standard applies to Samsø intended for direct consumption or for further processing in conformity with the description in Section 2 of this Standard.

2 DESCRIPTION

Samsø is a ripened hard cheese in conformity with the *General Standard for Cheese* (CXS 283-1978). The body has a near white or ivory through to light yellow or yellow colour and a firm-textured (when pressed by thumb) texture suitable for cutting, with few to plentiful, evenly distributed, smooth and round pea to cherry sized (or mostly up to 20 mm in diameter) gas holes, but few openings and splits are acceptable. The shape is a flat cylindrical, flat square or rectangular. The cheese is sold with or without[1] a hard, dry rind, which may be coated.

For Samsø ready for consumption, the ripening procedure to develop flavour and body characteristics is normally from 3 weeks at 8–17 °C depending on the extent of maturity required. Alternative ripening conditions (including the addition of ripening enhancing enzymes) may be used, provided the cheese exhibits similar physical, biochemical and sensory properties as those achieved by the previously stated ripening procedure. Samsø intended for further processing need not exhibit the same degree of ripening when justified through technical and/or trade needs.

3 ESSENTIAL COMPOSITION AND QUALITY FACTORS

3.1 Raw materials

Cows' milk or buffaloes' milk, or their mixtures, and products obtained from these milks.

3.2 Permitted ingredients

- Starter cultures of harmless lactic acid and/or flavour producing bacteria and cultures of other harmless micro-organisms;
- Rennet or other safe and suitable coagulating enzymes;
- Sodium chloride and potassium chloride as a salt substitute;
- Potable water;
- Safe and suitable enzymes to enhance the ripening process;

---

[1] This is not to mean that the rind has been removed before sale, instead the cheese has been ripened and/or kept in such a way that no rind is developed (a "rindless" cheese). Ripening film is used in the manufacture of rindless cheese. Ripening film may also constitute the coating that protects the cheese. For rindless cheese see also the Appendix to the *General Standard for Cheese* (CXS 283-1978).

- Safe and suitable processing aids;
- Rice, corn and potato flours and starches: Notwithstanding the provisions in the *General Standard for Cheese* (CXS 283-1978), these substances can be used in the same function as anti-caking agents for treatment of the surface of cut, sliced, and shredded products only, provided they are added only in amounts functionally necessary as governed by Good Manufacturing Practice, taking into account any use of the anti-caking agents listed in Section 4.

3.3 Composition

Milk constituent	Minimum content (m/m)	Maximum content (m/m)	Reference level (m/m)
Milk fat in dry matter	30%	Not restricted	45% to 55%
Dry matter	Depending on the fat in dry matter content, according to the table below		
	Fat in dry matter content (m/m)	Corresponding minimum dry matter content (m/m)	
	Equal to or above 30% but less than 40%	46%	
	Equal to or above 40% but less than 45%	52%	
	Equal to or above 45% but less than 55%	54%	
	Equal to or above 55%	59%	

Compositional modifications beyond the minima and maxima specified above for milk fat and dry matter are not considered to be in compliance with Section 4.3.3 of the *General Standard for the Use of Dairy Terms* (CXS 206-1999).

4 FOOD ADDITIVES

Only those additives classes indicated as justified in the table below may be used for the product categories specified. Anticaking agents, colours and preservatives used in accordance with Tables 1 and 2 of the *General Standard for Food Additives* (CXS 192-1995) in food category 01.6.2.1 (Ripened cheese, includes rind) and only certain acidity regulators and anticaking agents in Table 3 are acceptable for use in foods conforming to this standard.

Additive functional class	Justified use	
	Cheese mass	Surface/rind treatment
Colours	×[a]	–
Bleaching agents	–	–
Acidity regulators	×	–
Stabilizers	–	–
Thickeners	–	–
Emulsifiers	–	–
Antioxidants	–	–

Additive functional class	Justified use	
	Cheese mass	Surface/rind treatment
Preservatives	×	×
Foaming agents	–	–
Anti–caking agents	–	×(b)

(a) Only to obtain the colour characteristics, as described in Section 2.

(b) For the surface of sliced, cut, shredded or grated cheese, only.

× The use of additives belonging to the class is technologically justified.

– The use of additives belonging to the class is not technologically justified.

5 CONTAMINANTS

The products covered by this Standard shall comply with the Maximum Levels for contaminants that are specified for the product in the *General Standard for Contaminants and Toxins in Food and Feed* (CXS 193-1995).

The milk used in the manufacture of the products covered by this Standard shall comply with the Maximum Levels for contaminants and toxins specified for milk by the *General Standard for Contaminants and Toxins in Food and Feed* (CXS 193-1995) and with the maximum residue limits for veterinary drug residues and pesticides established for milk by the CAC.

6 HYGIENE

It is recommended that the product covered by the provisions of this standard be prepared and handled in accordance with the appropriate sections of the *General Principles of Food Hygiene* (CXC 1-1969), the *Code of Hygienic Practice for Milk and Milk Products* (CXC 57-2004) and other relevant Codex texts such as Codes of Hygienic Practice and Codes of Practice. The products should comply with any microbiological criteria established in accordance with the *Principles and Guidelines for the Establishment and Application of Microbiological Criteria Related to Foods* (CXG 21-1997).

7 LABELLING

In addition to the provisions of the *General Standard for the Labelling of Prepackaged Foods* (CXS 1-1985) and the *General Standard for the Use of Dairy Terms* (CXS 206-1999), the following specific provisions apply.

7.1 Name of the food

The name Samsø may be applied in accordance with section 4.1 of the *General Standard for the Labelling of Prepackaged Foods* (CXS 1-1985), provided that the product is in conformity with this Standard. Where customary in the country of retail sale, alternative spelling may be used.

The use of the name is an option that may be chosen only if the cheese complies with this standard. Where the name is not used for a cheese that complies with this standard, the naming provisions of the *General Standard for Cheese* (CXS 283-1978) apply.

The designation of products in which the fat content is below or above the reference

range but above the absolute minimum specified in section 3.3 of this Standard shall be accompanied by an appropriate qualification describing the modification made or the fat content (expressed as fat in dry matter or as percentage by mass whichever is acceptable in the country of retail sale), either as part of the name or in a prominent position in the same field of vision. Suitable qualifiers are the appropriate characterizing terms specified in Section 7.2 of the *General Standard for Cheese* (CXS 283-1978) or a nutritional claim in accordance with the *Guidelines for Use of Nutrition and Health Claims* (CXG 23-1997)[1].

The designation may also be used for cut, sliced, shredded or grated products made from cheese which cheese is in conformity with this Standard.

7.2 Country of origin

The country of origin (which means the country of manufacture, not the country in which the name originated) shall be declared. When the product undergoes substantial transformation[2] in a second country, the country in which the transformation is performed shall be considered to be the country of origin for the purpose of labelling.

7.3 Declaration of milk fat content

The milk fat content shall be declared in a manner found acceptable in the country of retail sale, either (i) as a percentage by mass, (ii) as a percentage of fat in dry matter, or (iii) in grams per serving as quantified in the label, provided that the number of servings is stated.

7.4 Date marking

Notwithstanding the provisions of Section 4.7.1 of the *General Standard for the Labelling of Prepackaged Foods* (CXS 1-1985), the date of manufacture may be declared instead of the minimum durability information, provided that the product is not intended to be purchased as such by the final consumer.

7.5 Labelling of non-retail containers

Information specified in Section 7 of this Standard and Sections 4.1 to 4.8 of the *General Standard for the Labelling of Prepackaged Foods* (CXS 1-1985) and, if necessary, storage instructions, shall be given either on the container or in accompanying documents, except that the name of the product, lot identification, and the name of the manufacturer or packer shall appear on the container, and in the absence of such a container, on the product itself. However, lot identification and the name and address may be replaced by an identification mark, provided that such mark is clearly identifiable with the accompanying documents.

8 METHODS OF SAMPLING AND ANALYSIS

For checking the compliance with this standard, the methods of analysis and sampling contained in the *Recommended Methods of Analysis and Sampling* (CXS 234-1999) relevant to the provisions in this standard, shall be used.

1　For the purpose of comparative nutritional claims, the minimum fat content of 45% fat in dry matter constitutes the reference.

2　For instance, repackaging, cutting, slicing, shredding and grating is not regarded as substantial transformation.

埃门塔尔干酪

STANDARD FOR EMMENTAL

CXS 269-1967

原为CODEX STAN C 9-1967 – 埃门塔尔干酪标准。1967年通过。2007年修订。2008年、2010年、2013年、2018年、2019年修正。

1 范围

本标准适用于符合本标准第2条所述即食或用于进一步加工的埃门塔尔干酪。

2 说明

埃门塔尔干酪是指符合《干酪通则》（CXS 283-1978）的成熟硬质干酪。干酪外观呈白色、银白色、淡黄色、橘黄色，其质地富有弹性、可切片但不粘连，并有数目不一、规则分布、或明或暗、樱桃及核桃大小（或者大多数直径在1～5 cm）的气孔，允许个别开口或裂口。埃门塔尔干酪通常为轮状或块状，重40 kg及以上，但个别国家也允许其他重量，只要干酪具备类似的物理、生化和感官特性。干酪加工及销售时带有或不带[1]坚硬、干燥外皮均可；其典型风味为清淡、似坚果、有甜味。

对于即食的埃门塔尔干酪，调制味道和外观特征的催熟程序在10～25 ℃下通常需要2个月，取决于所需成熟度。在遵守最低6周的时间，且干酪具备和前述催熟程序同样的物理、生化和感官特性情况下，可以使用其他催熟条件（包括添加加速熟化的酶）。如有合理技术或贸易需求，则进一步加工的埃门塔尔干酪无需具备同等成熟度。

3 基本成分和质量要求

3.1 基本原料

牛乳、水牛乳或它们的混合乳，以及源于这些乳的制品。

3.2 其他配料

— 无害乳酸发酵剂和/或产香菌和其他无害微生物的培养剂；

— 凝乳酶或其他安全、适宜的凝固酶；

— 作为代盐制品的氯化钠和氯化钾；

---

[1] 并非在销售前去掉外皮，相反干酪在成熟后，以不产生外皮方式保存（无皮干酪）。在无皮干酪加工过程中，使用催熟膜。催熟膜也可以成为保护干酪的外衣。无皮干酪也可参见《干酪通则》附件（CXS 283-1978）。

- 安全、适宜的加工助剂;
- 饮用水;
- 促进成熟的安全、适宜的酶;
- 大米、玉米、马铃薯粉和淀粉:尽管有《干酪通则》(CXS 283-1978),但这些物质同样可作为抗结块剂用作处理切块、切片和切碎产品表面,只要添加量是良好操作规范所要求达到期待功能的必需量,同时考虑第4条规定的抗结块剂的使用。

3.3 成分

乳成分	最低含量（m/m）	最高含量（m/m）	参考含量（m/m）
干物质乳脂	45%	/	45%~55%
干物质	根据干物质脂肪含量,参见下表		
	干物质脂肪含量（m/m）	相应最低干物质含量（m/m）	
	≥45%，<50%	60%	
	≥50%，<60%	62%	
	≥60%	67%	
待售干酪中丙酸[a]	≥150 mg/100 g		
钙含量[a]	≥800 mg/100 g		

(a) 这些标准旨在为下面两项验证（设计生产程序前的初步评估）提供依据:（i）意图采用的发酵和催熟条件是否能够达到培养菌的丙酸活动目的;（ii）凝乳管理和pH值制定是否能够实现质地特征。

乳脂和干物质含量超过规定的最大值和最小值被视为不符合《乳品术语》(CXS 206-1999)第4.3.3条。

3.4 主要生产特点

埃门塔尔干酪利用微生物发酵法生产,先用产菌的嗜热乳酸进行首次（乳糖）发酵,二次（乳酸）发酵以产菌的丙酸活动为主要特征。在温度显著超过[1]凝结温度后再加热凝乳。

4 食品添加剂

只有下表中允许的添加剂类别才可在特定产品类别中使用。根据《食品添加剂通用

---

1 本标准规定的达到构成和感官特征所需要的温度取决于许多其他技术因素,包括用于埃门塔尔干酪生产的牛乳的合适度、凝固酶的选择及其活动、主要和次要发酵剂的选择及活动、乳清排水的pH值、去除乳清的节点、催熟/储藏条件等。当地条件不同,这些其他因素也会不同:在很多情况下,特别是采用传统技术的情况下,烹饪温度一般在50℃左右;在其他情况下,可能会高于或低于此温度。

标准》（CXS 192-1995）表1和表2规定，在食品分类01.6.2.1（成熟干酪，包括外衣）中使用的抗结块剂、着色剂和防腐剂，以及表3中仅部分特定的酸度调节剂和抗结块剂可用于符合本标准的食品。

添加剂功能类别	合理使用	
	干酪量	表面/外皮处理
着色剂	×(a)	–
漂白剂	–	–
酸度调节剂	×	–
稳定剂	–	–
增稠剂	–	–
乳化剂	–	–
抗氧化剂	–	–
防腐剂	×	×
发泡剂	–	–
抗结块剂	–	×(a)

（a）仅为获得第2条所述颜色特征。
（b）仅用于切片、切块、切碎或磨碎的乳酪表面。
× 使用该类别添加剂符合技术要求。
– 使用该类别添加剂不符合技术要求。

5　污染物

本标准所涉及的产品应符合《食品及饲料中污染物和毒素通用标准》（CXS 193-1995）规定的污染物最大限量。

本标准所涉及的产品在加工中使用的牛乳应符合《食品及饲料中污染物和毒素通用标准》（CXS 193-1995）规定的乳中污染物和毒素最大限量以及国际食品法典委员会设定的农药和兽药最大残留限量。

6　卫生要求

建议本标准所涉及的产品应遵循《食品卫生总则》（CXC 1-1969）、《乳及乳制品卫生操作规范》（CXC 57-2004）以及《卫生操作规范》和《生产操作规范》等其他相关法典文本。产品应符合《食品微生物标准制定与实施原则和准则》（CXG 21-1997）规定的所有微生物标准。

7　标识

除符合《预包装食品标识通用标准》（CXS 1-1985）和《乳品术语》（CXS 206-1999）外，还应符合下列具体规定。

7.1 产品名称

埃门塔尔干酪名称应符合《预包装食品标识通用标准》（CXS 1-1985）第4.1条，只要产品符合该标准，产品销售国的习惯拼写也可被使用。

只要干酪符合本标准，就可选择使用此名称。对符合本标准但不使用此名称的干酪，应按《干酪通则》（CXS 283-1978）命名规定来命名。

对于脂肪含量超过本标准第3.3条规定的参考含量范围，但高于绝对最低含量的产品，在命名时须附上适当修饰语，作为名称的一部分或处于与名称同一视野内的显眼位置，说明所做调整或脂肪含量（以干物质中脂肪含量或质量百分比表示，视产品销售国的接受情况）。适当修饰语是指《干酪通则》（CXS 283-1978）第7.2条所规定的特征描述用语或是符合《营养和保健声明使用准则》（CXG 23-1997）[1]的营养声明。

该名称也可用于符合本标准规定的干酪做成的切块、切片、切碎或磨碎产品。

7.2 原产国

应当标记原产国（即生产国，而非名称起源国）。当产品在第二个国家有实质性改造[2]，则以标识目的进行改造的国家应视为原产国。

7.3 乳脂含量说明

乳脂含量应以销售国可接受的方式声明：（i）质量百分比形式；（ii）干物质中脂肪百分比形式；（iii）如果产品标识上标明了份数，也可用每份中乳脂重量（g）表示。

7.4 日期标识

尽管《预包装食品标识通用标准》（CXS 1-1985）第4.7.1条规定，仍应说明生产日期而非最低保存时间信息，只要产品并非直接出售给最终消费者。

7.5 非零售包装标识

在包装容器上除要标明产品名称、批次、生产厂家或包装商的名称和地址外，在包装容器上或附随说明书中，也应按本标准第7条和《预包装食品标识通用标准》（CXS 1-1985）第4.1～4.8条所要求的信息以及储存说明（在必要的地方）加以陈述。如果批次、生产厂家或包装商的名称和地址可以在附随的文件标明，则可以用一个识别标识来代替。

8 抽样和分析方法

为检查产品是否符合本标准，应采用《分析和抽样推荐性方法》（CXS 234-1999）中涉及本标准规定的分析和抽样方法。

---

1 为比较营养声明，干物质中脂肪含量最低参考值为45%。
2 例如，重新包装、切块、切片、切碎和磨碎不构成实质性改变。

附录

附 加 信 息

以下补充信息对前文各条款的规定不构成影响，前文规定内容对产品标识、食品名称的使用以及食品安全性至关重要。

1　外观特征

通常尺寸

形状	轮形	块形
高度	12～30 cm	12～30 cm
直径	70～100 cm	—
最低重量	60 kg	40 kg

2　生产方法

2.1　发酵工艺：微生物产酸发酵。

STANDARD FOR EMMENTAL

CXS 269-1967

Formerly CODEX STAN C 9-1967 – Standard for Emmentaler.

Adopted in 1967. Revised in 2007. Amended in 2008, 2010, 2013, 2018, 2019.

1 SCOPE

This Standard applies to Emmental intended for direct consumption or for further processing in conformity with the description in Section 2 of this Standard.

2 DESCRIPTION

Emmental is a ripened hard cheese in conformity with the *General Standard for Cheese* (CXS 283-1978). The body has a ivory through to light yellow or yellow colour and an elastic, sliceable but not sticky texture, with regular, scarce to plentiful distributed, mat to brilliant, cherry to walnut sized (or mostly from 1 to 5 cm in diameter) gas holes, but few openings and splits are acceptable. Emmental is typically manufactured as wheels and blocks of weights from 40 kg or more but individual countries may on their territory permit other weights provided that the cheese exhibit similar physical, biochemical and sensory properties. The cheese is manufactured and sold with or without[1] a hard, dry rind. The typical flavour is mild, nut-like and sweet, more or less pronounced.

For Emmental ready for consumption, the ripening procedure to develop flavour and body characteristics is normally from 2 months at 10–25 °C depending on the extent of maturity required. Alternative ripening conditions (including the addition of ripening enhancing enzymes) may be used, provided a minimum period of 6 weeks is observed and provided the cheese exhibits similar physical, biochemical and sensory properties as those achieved by the previously stated ripening procedure. Emmental intended for further processing need not exhibit the same degree of ripening, when justified through technical and/or trade needs.

3 ESSENTIAL COMPOSITION AND QUALITY FACTORS

3.1 Raw materials

Cows' milk or buffaloes' milk, or their mixtures, and products obtained from these milks.

3.2 Permitted ingredients

- Starter cultures of harmless lactic acid and/or flavour producing bacteria and cultures of other harmless micro-organisms;

---

[1] This is not to mean that the rind has been removed before sale, instead the cheese has been ripened and/or kept in such a way that no rind is developed (a "rindless" cheese). Ripening film is used in the manufacture of rindless cheese. Ripening film may also constitute the coating that protects the cheese. For rindless cheese see also the Appendix to the *General Standard for Cheese* (CXS 283-1978).

- Rennet or other safe and suitable coagulating enzymes;
- Sodium chloride and potassium chloride as a salt substitute;
- Safe and suitable processing aids;
- Potable water;
- Safe and suitable enzymes to enhance the ripening process;
- Rice, corn and potato flours and starches: Notwithstanding the provisions in the *General Standard for Cheese* (CXS 283-1978), these substances can be used in the same function as anti-caking agents for treatment of the surface of cut, sliced, and shredded products only, provided they are added only in amounts functionally necessary as governed by Good Manufacturing Practice, taking into account any use of the anti-caking agents listed in Section 4.

3.3 Composition

Milk constituent	Minimum content (m/m)	Maximum content (m/m)	Reference level (m/m)
Milk fat in dry matter	45%	Not restricted	45% to 55%
Dry matter	Depending on the fat in dry matter content, according to the table below		
	Fat in dry matter content (m/m)		Corresponding minimum dry matter content (m/m)
	Equal to or above 45% but less than 50%		60%
	Equal to or above 50% but less than 60%		62%
	Equal to or above 60%		67%
Propionic acid in cheese ready for sale[a]	minimum 150 mg/100g		
Calcium content[a]	minimum 800 mg/100g		

(a) The purpose of these criteria is to provide targets for the validation (initial assessment prior to the design of the manufacturing process), respectively, of (i) whether the intended fermentation and ripening conditions are capable of achieving the activity of propionic acid producing bacteria, and of (ii) whether the curd management and pH development are capable of obtaining the characteristic texture.

Compositional modifications beyond the minima and maxima specified above for milk fat and dry matter are not considered to be in compliance with section 4.3.3 of the *General Standard for the Use of Dairy Terms* (CXS 206-1999).

3.4 Essential manufacturing characteristics

Emmental is obtained by microbiological fermentation, using thermophilic lactic acid producing bacteria for the primary (lactose) fermentation; the secondary (lactate) fermentation is characterized by the activity of propionic acid producing bacteria. The curd is heated after cutting to a temperature significantly above[1] the coagulation temperature.

4 FOOD ADDITIVES

Only those additives classes indicated as justified in the table below may be used for the product categories specified. Anticaking agents, colours and preservatives used in accordance with Tables 1 and 2 of the *General Standard for Food Additives* (CXS 192-1995) in food category 01.6.2.1 (Ripened cheese, includes rind) and only certain acidity regulators and anticaking agents in Table 3 are acceptable for use in foods conforming to this standard.

Additive functional class:	Justified use	
	Cheese mass	Surface/rind treatment
Colours	×(a)	–
Bleaching agents	–	–
Acidity regulators	×	–
Stabilizers	–	–
Thickeners	–	–
Emulsifiers	–	–
Antioxidants	–	–
Preservatives	×	×
Foaming agents	–	–
Anti–caking agents	–	×(b)

(a) Only to obtain the colour characteristics, as described in Section 2.
(b) For the surface of sliced, cut, shredded or grated cheese, only.
× The use of additives belonging to the class is technologically justified.
– The use of additives belonging to the class is not technologically justified.

5 CONTAMINANTS

The products covered by this Standard shall comply with the Maximum Levels for contaminants that are specified for the product in the *General Standard for Contaminants and Toxins in Food and Feed* (CXS 193-1995).

---

1 The temperature required to obtain the compositional and sensory characteristics specified by this Standard depends on a number of other technology factors, including the suitability of the milk for Emmental manufacture, the choice and activity of coagulating enzymes and of primary and secondary starter cultures, the pH at whey drainage and at the point of whey removal, and the ripening/storage conditions. These other factors differ according to local circumstances: In many cases, in particular where traditional technology is applied, a cooking temperatures of approx. 50 °C is typically applied; In other cases, temperatures above and below are applied.

The milk used in the manufacture of the products covered by this Standard shall comply with the Maximum Levels for contaminants and toxins specified for milk by the *General Standard for Contaminants and Toxins in Food and Feed* (CXS 193-1995) and with the maximum residue limits for veterinary drug residues and pesticides established for milk by the CAC.

6 HYGIENE

It is recommended that the product covered by the provisions of this standard be prepared and handled in accordance with the appropriate sections of the *General Principles of Food Hygiene* (CXC 1-1969), the *Code of Hygienic Practice for Milk and Milk Products* (CXC 57-2004) and other relevant Codex texts such as Codes of Hygienic Practice and Codes of Practice. The products should comply with any microbiological criteria established in accordance with the *Principles and Guidelines for the Establishment and Application of Microbiological Criteria Related to Foods* (CXG 21-1997).

7 LABELLING

In addition to the provisions of the *General Standard for the Labelling of Prepackaged Foods* (CXS 1-1985) and the *General Standard for the Use of Dairy Terms* (CXS 206-1999), the following specific provisions apply.

7.1 Name of the food

The names Emmental or Emmentaler may be applied in accordance with Section 4.1 of the *General Standard for the Labelling of Prepackaged Foods* (CXS 1-1985), provided that the product is in conformity with this Standard. Where customary in the country of retail sale, alternative spelling may be used.

The use of the name is an option that may be chosen only if the cheese complies with this standard. Where the name is not used for a cheese that complies with this standard, the naming provisions of the *General Standard for Cheese* (CXS 283-1978) apply.

The designation of products in which the fat content is above the reference range specified in section 3.3 of this Standard shall be accompanied by an appropriate qualification describing the modification made or the fat content (expressed as fat in dry matter or as percentage by mass whichever is acceptable in the country of retail sale), either as part of the name or in a prominent position in the same field of vision. Suitable qualifiers are the appropriate characterizing terms specified in Section 7.2 of the *General Standard for Cheese* (CXS 283-1978) or a nutritional claim in accordance with the *Guidelines for Use of Nutrition and Health Claims* (CXG 23-1997)[1].

The designation may also be used for cut, sliced, shredded or grated products made from cheese which cheese is in conformity with this Standard.

7.2 Country of origin

The country of origin (which means the country of manufacture, not the country in which the name originated) shall be declared. When the product undergoes substantial transformation[2] in a second country, the country in which the transformation is performed shall be

---

[1] For the purpose of comparative nutritional claims, the minimum fat content of 45% fat in dry matter constitutes the reference.

[2] For instance, repackaging, cutting, slicing, shredding and grating is not regarded as substantial transformation.

considered to be the country of origin for the purpose of labelling.

7.3 Declaration of milk fat content

The milk fat content shall be declared in a manner found acceptable in the country of retail sale. either (i) as a percentage by mass, (ii) as a percentage of fat in dry matter, or (iii) in grams per serving as quantified in the label, provided that the number of servings is stated.

7.4 Date marking

Notwithstanding the provisions of Section 4.7.1 of the *General Standard for the Labelling of Prepackaged Foods* (CXS 1-1985), the date of manufacture may be declared instead of the minimum durability information, provided that the product is not intended to be purchased as such by the final consumer.

7.5 Labelling of non-retail containers

Information specified in Section 7 of this Standard and Sections 4.1 to 4.8 of the *General Standard for the Labelling of Prepackaged Foods* (CXS 1-1985) and, if necessary, storage instructions, shall be given either on the container or in accompanying documents, except that the name of the product, lot identification, and the name of the manufacturer or packer shall appear on the container, and in the absence of such a container, on the product itself. However, lot identification and the name and address may be replaced by an identification mark, provided that such mark is clearly identifiable with the accompanying documents.

8 METHODS OF SAMPLING AND ANALYSIS

For checking the compliance with this standard, the methods of analysis and sampling contained in the *Recommended Methods of Analysis and Sampling* (CXS 234-1999) relevant to the provisions in this standard, shall be used.

STANDARD FOR EMMENTAL

APPENDIX

ADDITIONAL INFORMATION

The additional information below does not affect the provisions in the preceding sections which are those that are essential to the product identity, the use of the name of the food and the safety of the food.

1 **Appearance characteristics**

Usual dimensions

Shape	Wheel	Block
Height	12–30 cm	12–30 cm
Diameter	70–100 cm	–
Minimum weight	60 kg	40 kg

2 **Method of manufacture**

2.1 Fermentation procedure: Microbiologically derived acid development.

泰尔西特干酪

STANDARD FOR TILSITER

CXS 270-1968

原为CODEX STAN C-11-1968。1968年通过。2007年修订。
2008年、2010年、2013年、2018年、2019年修正。

1 范围

本标准适用于符合标准第2条所述即食或用于进一步加工的泰尔西特干酪。

2 说明

泰尔西特干酪是指符合《干酪通则》（CXS 283-1978）的成熟硬质或半硬质干酪。干酪外观呈白色、银白色、淡黄色、橘黄色，其质地坚硬（用拇指摁压）、适合切割，有少数或多个均匀分布、光滑的气孔。干酪加工及销售时带或不带非常干燥、附有涂层的外皮均可，也可以有外衣[1]。

对于即食的泰尔西特干酪，调制味道和外观特征的催熟程序在10～16 ℃下通常需要3周，取决于所需成熟度。在干酪具备和前述催熟程序同样的物理、生化和感官特性的条件下，可以使用其他催熟条件（包括添加加速熟化的酶）。如有合理技术或贸易需求，则进一步加工的泰尔西特干酪无需具备同等成熟度。

3 基本成分和质量要求

3.1 基本原料

牛乳、水牛乳或其混合乳，以及源于这些乳的制品。

3.2 其他配料

- 无害乳酸发酵剂和/或产香菌和其他无害微生物的培养剂；
- 凝乳酶或其他安全、适宜的凝固酶；
- 作为代盐制品的氯化钠和氯化钾；
- 饮用水；
- 促进成熟的安全、适宜的酶；
- 安全、适宜的加工助剂；

---

[1] 并非在销售前去掉外皮，相反干酪催熟后，以不产生外皮方式保存（无皮干酪）。在无皮干酪加工过程中，使用催熟膜。催熟膜也可以成为保护干酪的外衣。无皮干酪也可参见《干酪通则》（CXS 283-1978）附件。

— 大米、玉米、马铃薯粉和淀粉：尽管有《干酪通则》（CXS 283-1978），但这些物质同样可作为抗结块剂处理切块、切片和切碎产品表面，只要添加量是良好操作规范所要求达到期待功能的必需量，同时考虑第4条规定的抗结块剂的使用。

3.3 成分

乳成分	最低含量（m/m）	最高含量（m/m）	参考含量（m/m）
干物质乳脂	30%	/	45%～55%
干物质	根据干物质脂肪含量，参见下表		
	≥30%，<40%		49%
	≥40%，<45%		53%
	≥45%，<50%		55%
	≥50%，<60%		57%
	≥60%，<85%		61%

乳脂和干物质含量超过规定的最大值和最小值被视为不符合《乳品术语》（CXS 206-1999）第4.3.3条。

4 食品添加剂

只有下表中标注为合理使用的添加剂类别才可以用于指定产品类别。根据《食品添加剂通用标准》（CXS 192-1995）表1和表2在食品类别01.6.2.1（成熟干酪，包括外皮）中使用的抗结块剂、着色剂和防腐剂，以及表3中仅部分酸度调节剂和抗结块剂可用于符合本标准的食品。

添加剂功能类别	合理使用	
	干酪量	表面/表皮处理
着色剂	×[a]	-
漂白剂	-	-
酸度调节剂	×	-
稳定剂	-	-
增稠剂	-	-
乳化剂	-	-
抗氧化剂	-	-
防腐剂	×	×

添加剂功能类别	合理使用	
	干酪量	表面/表皮处理
发泡剂	-	-
抗结块剂	-	×(b)

（a）仅为获得第2条所述颜色特征。
（b）仅用于切片、切块、切碎或磨碎的乳酪表面。
× 使用该类别添加剂符合技术要求。
- 使用该类别添加剂不符合技术要求。

5 污染物

本标准所涉及的产品应符合《食品及饲料中污染物和毒素通用标准》（CXS 193-1995）规定的污染物最大限量。

本标准所涉及的产品在加工中使用的牛乳应符合《食品及饲料中污染物和毒素通用标准》（CXS 193-1995）规定的乳中污染物和毒素最大限量以及国际食品法典委员会设定的农药和兽药最大残留限量。

6 卫生要求

建议本标准所涉及的产品应遵循《食品卫生总则》（CXC 1-1969）、《乳及乳制品卫生操作规范》（CXC 57-2004）以及《卫生操作规范》和《生产操作规范》等其他相关法典文本。产品应符合《食品微生物标准制定与实施原则和准则》（CXG 21-1997）规定的所有微生物标准。

7 标识

除符合《预包装食品标识通用标准》（CXS 1-1985）和《乳品术语》（CXS 206-1999）外，还应符合下列具体规定。

7.1 产品名称

泰尔西特干酪名称应符合《预包装食品标识通用标准》（CXS 1-1985）第4.1条，只要产品符合该标准，产品销售国的习惯拼写也可被使用。

只要干酪符合本标准，就可选择使用此名称。对符合本标准但不使用此名称的干酪，应按《干酪通则》（CXS 283-1978）命名规定来命名。

对于脂肪含量超过本标准第3.3条规定的参考含量范围，但高于绝对最低含量的产品，在命名时须附上适当修饰语，作为名称的一部分或处于与名称同一视野内的显眼位置，说明所做调整或脂肪含量（以干物质中脂肪含量或质量百分比表示，视产品销售国的接受情况）。适当修饰语是指《干酪通则》（CXS 283-1978）第7.2条所规定的特征描述用语或是符合《营养和保健声明使用准则》（CXG 23-1997）[1]的营养声明。

该名称也可用于符合本标准规定的干酪做成的切块、切片、切碎或磨碎产品。

7.2 原产国

应当标记原产国（生产国，而非名称起源国）。当产品在第二个国家有实质性改造[2]，则以标识目的进行改造的国家应视为原产国。

7.3 乳脂含量说明

乳脂含量应以销售国可接受的方式声明：（i）质量百分比形式；（ii）干物质中脂肪百分比形式；（iii）如果产品标识上标明了份数，也可用每份中乳脂重量（g）表示。

7.4 日期标识

尽管《预包装食品标识通用标准》（CXS 1-1985）第4.7.1条规定，仍应说明生产日期而非最低保存时间信息，只要产品并非直接出售给最终消费者。

7.5 非零售包装标识

在包装容器上除要标明产品名称、批次、生产厂家或包装商的名称和地址外，在包装容器上或附随说明书中，也应按本标准第7条和《预包装食品标识通用标准》（CXS 1-1985）第4.1~4.8条所要求的信息以及储存说明（在必要的地方）加以陈述。如果批次、生产厂家或包装商的名称和地址可以在附随的文件标明，则可以用一个识别标识来代替。

8 抽样和分析方法

为检查产品是否符合本标准，应采用《分析和抽样推荐性方法》（CXS 234-1999）中涉及本标准规定的分析和抽样方法。

---

[1] 为比较营养声明，干物质中脂肪含量最低参考值为45%。
[2] 例如，重新包装、切块、切片、切碎和磨碎不构成实质性改变。

STANDARD FOR TILSITER

CXS 270-1968

Formerly CODEX STAN C-11-1968. Adopted in 1968. Revised in 2007.

Amended in 2008, 2010, 2013, 2018, 2019.

1 SCOPE

This Standard applies to Tilsiter intended for direct consumption or for further processing in conformity with the description in Section 2 of this Standard.

2 DESCRIPTION

Tilsiter is a ripened firm/semi-hard cheese in conformity with the *General Standard for Cheese* (CXS 283-1978). The body has a near white or ivory through to light yellow or yellow colour and a firm-textured (when pressed by thumb) texture suitable for cutting, with irregularly shaped, shiny and evenly distributed gas holes. The cheese is manufactured and sold with or without[1] a well-dried smear-developed rind, which may be coated.

For Tilsiter ready for consumption, the ripening procedure to develop flavour and body characteristics is normally from 3 weeks at 10–16 °C depending on the extent of maturity required. Alternative ripening conditions (including the addition of ripening enhancing enzymes) may be used, provided the cheese exhibits similar physical, biochemical and sensory properties as those achieved by the previously stated ripening procedure. Tilsiter intended for further processing need not exhibit the same degree of ripening when justified through technical and/or trade needs.

3 ESSENTIAL COMPOSITION AND QUALITY FACTORS

3.1 Raw materials

Cows' milk or buffaloes' milk, or their mixtures, and products obtained from these milks.

3.2 Permitted ingredients

- Starter cultures of harmless lactic acid and/or flavour producing bacteria and cultures of other harmless micro-organisms;
- Rennet or other safe and suitable coagulating enzymes;
- Sodium chloride and potassium chloride as a salt substitute;
- Potable water;
- Safe and suitable enzymes to enhance the ripening process;
- Safe and suitable processing aids;

---

1 This is not to mean that the rind has been removed before sale, instead the cheese has been ripened and/or kept in such a way that no rind is developed (a "rindless" cheese). Ripening film is used in the manufacture of rindless cheese. Ripening film may also constitute the coating that protects the cheese. For rindless cheese, see also the Appendix to the *General Standard for Cheese* (CXS 283-1978).

– Rice, corn and potato flours and starches: Notwithstanding the provisions in the *General Standard for Cheese* (CXS 283-1978), these substances can be used in the same function as anti-caking agents for treatment of the surface of cut, sliced, and shredded products only, provided they are added only in amounts functionally necessary as governed by Good Manufacturing Practice, taking into account any use of the anti-caking agents listed in Section 4.

3.3 Composition

Milk constituent	Minimum content (m/m)	Maximum content (m/m)	Reference level (m/m)
Milk fat in dry matter	30%	Not restricted	45% to 55%
Dry matter	Depending on the fat in dry matter content, according to the table below		
	Fat in dry matter content (m/m)		Corresponding minimum dry matter content (m/m)
	Equal to or above 30% but less than 40%		49%
	Equal to or above 40% but less than 45%		53%
	Equal to or above 45% but less than 50%		55%
	Equal to or above 50% but less than 60%		57%
	Equal to or above 60% but less than 85%		61%

Compositional modifications beyond the minima and maxima specified above for milk fat and dry matter are not considered to be in compliance with section 4.3.3 of the *General Standard for the Use of Dairy Terms* (CXS 206-1999).

4 FOOD ADDITIVES

Only those additives classes indicated as justified in the table below may be used for the product categories specified. Anticaking agents, colours and preservatives used in accordance with Tables 1 and 2 of the *General Standard for Food Additives* (CXS 192-1995) in food category 01.6.2.1 (Ripened cheese, includes rind) and only certain acidity regulators and anticaking agents in Table 3 are acceptable for use in foods conforming to this standard.

Additive functional class	Justified use	
	Cheese mass	Surface/rind treatment
Colours	×[a]	–
Bleaching agents	–	–
Acidity regulators	×	–
Stabilizers	–	–
Thickeners	–	–
Emulsifiers	–	–

Additive functional class	Justified use	
	Cheese mass	Surface/rind treatment
Antioxidants	–	–
Preservatives	×	×
Foaming agents	–	–
Anti–caking agents	–	×(b)

(a) Only to obtain the colour characteristics, as described in Section 2.

(b) For the surface of sliced, cut, shredded or grated cheese, only.

× The use of additives belonging to the class is technologically justified.

– The use of additives belonging to the class is not technologically justified.

5 CONTAMINANTS

The products covered by this Standard shall comply with the Maximum Levels for contaminants that are specified for the product in the *General Standard for Contaminants and Toxins in Food and Feed* (CXS 193-1995).

The milk used in the manufacture of the products covered by this Standard shall comply with the Maximum Levels for contaminants and toxins specified for milk by the *General Standard for Contaminants and Toxins in Food and Feed* (CXS 193-1995) and with the maximum residue limits for veterinary drug residues and pesticides established for milk by the CAC.

6 HYGIENE

It is recommended that the product covered by the provisions of this standard be prepared and handled in accordance with the appropriate sections of the *General Principles of Food Hygiene* (CXC 1-1969), the *Code of Hygienic Practice for Milk and Milk Products* (CXC 57-2004) and other relevant Codex texts such as Codes of Hygienic Practice and Codes of Practice. The products should comply with any microbiological criteria established in accordance with the *Principles and Guidelines for the Establishment and Application of Microbiological Criteria Related to Foods* (CXG 21-1997).

7 LABELLING

In addition to the provisions of the *General Standard for the Labelling of Prepackaged Foods* (CXS 1-1985) and the *General Standard for the Use of Dairy Terms* (CXS 206-1999), the following specific provisions apply.

7.1 Name of the food

The name Tilsiter may be applied in accordance with section 4.1 of the *General Standard for the Labelling of Prepackaged Foods* (CXS 1-1985), provided that the product is in conformity with this Standard. Where customary in the country of retail sale, alternative spelling may be used.

The use of the name is an option that may be chosen only if the cheese complies with this standard. Where the name is not used for a cheese that complies with this standard, the naming provisions of the *General Standard for Cheese* (CXS 283-1978) apply.

The designation of products in which the fat content is below or above the reference range but above the absolute minimum specified in section 3.3 of this Standard shall be accompanied by an appropriate qualification describing the modification made or the fat content (expressed as fat in dry matter or as percentage by mass whichever is acceptable in the country of retail sale), either as part of the name or in a prominent position in the same field of vision. Suitable qualifiers are the appropriate characterizing terms specified in Section 7.2 of the *General Standard for Cheese* (CXS 283-1978) or a nutritional claim in accordance with the *Guidelines for Use of Nutrition and Health Claims* (CXG 23-1997).[1]

The designation may also be used for cut, sliced, shredded or grated products made from cheese which cheese is in conformity with this Standard.

7.2 Country of origin

The country of origin (which means the country of manufacture, not the country in which the name originated) shall be declared. When the product undergoes substantial transformation[2] in a second country, the country in which the transformation is performed shall be considered to be the country of origin for the purpose of labelling.

7.3 Declaration of milk fat content

The milk fat content shall be declared in a manner found acceptable in the country of retail sale either (i) as a percentage by mass, (ii) as a percentage of fat in dry matter, or (iii) in grams per serving as quantified in the label, provided that the number of servings is stated.

7.4 Date marking

Notwithstanding the provisions of Section 4.7.1 of the *General Standard for the Labelling of Prepackaged Foods* (CXS 1-1985), the date of manufacture may be declared instead of the minimum durability information, provided that the product is not intended to be purchased as such by the final consumer.

7.5 Labelling of non-retail containers

Information specified in Section 7 of this Standard and Sections 4.1 to 4.8 of the *General Standard for the Labelling of Prepackaged Foods* (CXS 1-1985) and, if necessary, storage instructions, shall be given either on the container or in accompanying documents, except that the name of the product, lot identification, and the name of the manufacturer or packer shall appear on the container, and in the absence of such a container, on the product itself. However, lot identification and the name and address may be replaced by an identification mark, provided that such mark is clearly identifiable with the accompanying documents.

8 METHODS OF SAMPLING AND ANALYSIS

For checking the compliance with this standard, the methods of analysis and sampling contained in the *Recommended Methods of Analysis and Sampling* (CXS 234-1999) relevant to the provisions in this standard, shall be used.

---

1   For the purpose of comparative nutritional claims, the minimum fat content of 45% fat in dry matter constitutes the reference.

2   For instance, repackaging, cutting, slicing, shredding and grating is not regarded as substantial transformation.

圣宝林干酪

STANDARD FOR SAINT-PAULIN

CXS 271-1968

原为CODEX STAN C-13-1968。1968年通过。2007年修订。
2008年、2010年、2013年、2018年、2019年修正。

1　范围

本标准适用于符合本标准第2条所述即食或用于进一步加工的圣宝林干酪。

2　说明

圣宝林干酪是指符合《干酪通则》（CXS 283-1978）的成熟硬质或半硬质干酪。干酪外观呈白色、银白色、淡黄色、橘黄色，其质地坚硬（用拇指摁压）且富有弹性。一般无气孔，但允许个别开口或裂口。干酪加工及销售时带或不带干燥外皮或轻微湿润的外皮均可[1]。

对于即食的圣宝林干酪，调制味道和外观特征的催熟程序在10～17 ℃下通常需要1周，取决于所需成熟度。在干酪具备和前述催熟程序同样的物理、生化和感官特性的情况下，可以使用其他催熟条件（包括添加加速成熟的酶）。如有合理技术或贸易需求，用于进一步加工的圣宝林干酪无需具备同等成熟度。

3　基本成分和质量要求

3.1　基本原料

牛乳、水牛乳或它们的混合乳，以及源于这些乳的制品。

3.2　其他配料

- 无害乳酸发酵剂和/或产香菌和其他无害微生物的培养剂；
- 凝乳酶或其他安全、适宜的凝固酶；
- 作为代盐制品的氯化钠和氯化钾；
- 饮用水；
- 促进成熟的安全、适宜的酶；
- 安全、适宜的加工助剂；

---

[1] 并非在销售前去掉外皮，相反干酪在成熟后，以不产生外皮方式保存（无皮干酪）。在无皮干酪加工过程中，使用催熟膜。催熟膜也可以成为保护干酪的外衣。无皮干酪也可参见《干酪通则》附件（CXS 283-1978）。

— 大米、玉米、马铃薯粉和淀粉：尽管有《干酪通则》（CXS 283-1978），但这些物质同样可以作为抗结块剂处理切块、切片和切碎产品表面，只要添加量是良好操作规范所要求达到期待功能的必需量，同时考虑第4条规定的抗结块剂的使用。

3.3 成分

乳成分	最低含量（m/m）	最高含量（m/m）	参考含量（m/m）
干物质乳脂	40%	/	40%～50%
干物质	根据干物质脂肪含量，参见下表		
	干物质脂肪含量（m/m）	相应最低干物质含量（m/m）	
	≥40%，<60%	44%	
	≥60%	54%	

乳脂和干物质含量超过规定的最大值和最小值被视为不符合《乳品术语》（CXS 206-1999）第4.3.3条。

4 食品添加剂

只有下表中标注为合理使用的添加剂类别才可以用于指定产品类别。根据《食品添加剂通用标准》（CXS 192-1995）表1和表2在食品类别01.6.2.1（成熟干酪，包括外皮）中使用的抗结块剂、着色剂和防腐剂，以及表3中仅部分酸度调节剂和抗结块剂可用于符合本标准的食品。

添加剂功能类别	合理使用	
	干酪量	表面/外皮处理
着色剂	×(a)	–
漂白剂	–	–
酸度调节剂	×	–
稳定剂	–	–
增稠剂	–	–
乳化剂	–	–
抗氧化剂	–	–
防腐剂	×	×
发泡剂	–	–
抗结块剂	–	×(b)

（a）仅为获得第2条所述颜色特征。
（b）仅用于切片、切块、切碎或磨碎的乳酪表面。
× 使用该类别添加剂符合技术要求。
– 使用该类别添加剂不符合技术要求。

5　污染物

本标准所涉及的产品应符合《食品及饲料中污染物和毒素通用标准》（CXS 193-1995）规定的污染物最大限量。

本标准所涉及的产品在加工中使用的牛乳应符合《食品及饲料中污染物和毒素通用标准》（CXS 193-1995）规定的乳中污染物和毒素最大限量以及国际食品法典委员会设定的农药和兽药最大残留限量。

6　卫生要求

建议本标准所涉及的产品应遵循《食品卫生总则》（CXC 1-1969）、《乳及乳制品卫生操作规范》（CXC 57-2004）以及《卫生操作规范》和《生产操作规范》等其他相关法典文本。产品应符合《食品微生物标准制定与实施原则和准则》（CXG 21-1997）规定的所有微生物标准。

7　标识

除符合《预包装食品标识通用标准》（CXS 1-1985）和《乳品术语》（CXS 206-1999）外，还应符合下列具体规定。

7.1　产品名称

圣宝林干酪名称应符合《预包装食品标识通用标准》（CXS 1-1985）第4.1条，只要产品符合该标准，产品销售国的习惯拼写也可被使用。

只要干酪符合本标准，就可选择使用此名称。对符合本标准但不使用此名称的干酪，应按《干酪通则》（CXS 283-1978）命名规定来命名。

对于脂肪含量超过本标准第3.3条规定的参考含量范围，但高于绝对最低含量的产品，在命名时须附上适当修饰语，作为名称的一部分或处于与名称同一视野内的显眼位置，说明所做调整或脂肪含量（以干物质中脂肪含量或质量百分比表示，视产品销售国的接受情况）。适当修饰语是指《干酪通则》（CXS 283-1978）第7.2条所规定的特征描述用语或是符合《营养和保健声明使用准则》（CXG 23-1997）[1]的营养声明。

该名称也可用于符合本标准规定的干酪做成的切块、切片、切碎或磨碎产品。

7.2　原产国

应当标记原产国（即生产国，而非名称起源国）。当产品在第二个国家有实质性改造[2]，则以标识目的进行改造的国家应视为原产国。

---

[1]　为比较营养声明，干物质中脂肪含量最低参考值为40%。
[2]　例如，重新包装、切块、切片、切碎和磨碎不构成实质性改变。

7.3 乳脂含量说明

乳脂含量应以销售国可接受的方式声明：（i）质量百分比形式；（ii）干物质中脂肪百分比形式；（iii）如果产品标识上标明了份数，也可用每份中乳脂重量（g）表示。

7.4 日期标识

尽管有《预包装食品标识通用标准》（CXS 1-1985）第4.7.1条的规定，仍应说明生产日期而非最低保存时间信息，只要产品并非直接出售给最终消费者。

7.5 非零售包装标识

在包装容器上除要标明产品名称、批次、生产厂家或包装商的名称和地址外，在包装容器上或附随说明书中，也应按本标准第7条和《预包装食品标识通用标准》（CXS 1-1985）第4.1~4.8条所要求的信息以及储存说明（在必要的地方）加以陈述。如果批次、生产厂家或包装商的名称和地址可以在附随的文件标明，则可以用一个识别标识来代替。

8 抽样和分析方法

为检查产品是否符合本标准，应采用《分析和抽样推荐性方法》（CXS 234-1999）中涉及本标准规定的分析和抽样方法。

附录

附加信息

以下补充信息对前文各条款的规定不构成影响，前文规定内容对产品标识、食品名称的使用以及食品安全性至关重要。

1 外观特征

1.1 形状：略凸边的小圆柱体，其他形状也有可能。

1.2 尺寸和重量：

（a）通常变体：直径大约20 cm，最小重量为1.3 kg。

（b）"小圣宝林干酪"：直径8～13 cm，最小重量为150 g。

（c）"最小圣宝林干酪"：最小重量为20 g。

2 加工方法

2.1 发酵工艺：微生物产酸发酵。

2.2 其他特征：干酪在盐水中腌制。

3 限定

当干酪遵从上述关于直径和重量的规定时需要指定"小圣宝林干酪"和"最小圣宝林干酪"名称。

STANDARD FOR SAINT-PAULIN

CXS 271-1968

Formerly CODEX STAN C-13-1968. Adopted in 1968. Revised in 2007.

Amended in 2008, 2010, 2013, 2018, 2019.

1 SCOPE

This Standard applies to Saint-Paulin intended for direct consumption or for further processing in conformity with the description in Section 2 of this Standard.

2 DESCRIPTION

Saint-Paulin is a ripened firm/semi-hard cheese in conformity with the *General Standard for Cheese* (CXS 283-1978). The body has a near white or ivory through to light yellow or yellow colour and a firm-textured (when pressed by thumb) but flexible texture. Gas holes are generally absent, but few openings and splits are acceptable. The cheese is manufactured and sold with or without[1] a dry or slightly moist rind, which is hard, but elastic under thumb pressure, and which may be coated.

For Saint-Paulin ready for consumption, the ripening procedure to develop flavour and body characteristics is normally from 1 week at 10–17 °C depending on the extent of maturity required. Alternative ripening conditions (including the addition of ripening enhancing enzymes)may be used, provided the cheese exhibits similar physical, biochemical and sensory properties as those achieved by the previously stated ripening procedure. Saint-Paulin intended for further processing need not exhibit the same degree of ripening when justified through technical and/or trade needs.

3 ESSENTIAL COMPOSITION AND QUALITY FACTORS

3.1 Raw materials

Cows' milk or buffaloes' milk, or their mixtures, and products obtained from these milks.

3.2 Permitted ingredients

- Starter cultures of harmless lactic acid and/or flavour producing bacteria and cultures of other harmless micro-organisms;
- Rennet or other safe and suitable coagulating enzymes;
- Sodium chloride and potassium chloride as a salt substitute;
- Potable water;
- Safe and suitable enzymes to enhance the ripening process;
- Safe and suitable processing aids;

---

[1] This is not to mean that the rind has been removed before sale, instead the cheese has been ripened and/or kept in such a way that no rind is developed (a "rindless" cheese). Ripening film may be used in the manufacture of rindless cheese. Ripening film may also constitute the coating that protects the cheese. For rindless cheese, see also the Appendix to the *General Standard for Cheese* (CXS 283-1978).

– Rice, corn and potato flours and starches: Notwithstanding the provisions in the *General Standard for Cheese* (CXS 283-1978), these substances can be used in the same function as anti-caking agents for treatment of the surface of cut, sliced, and shredded products only, provided they are added only in amounts functionally necessary as governed by Good Manufacturing Practice, taking into account any use of the anti-caking agents listed in Section 4.

3.3 Composition

Milk constituent	Minimum content (m/m)	Maximum content (m/m)	Reference level (m/m)
Milkfat in dry matter	40%	Not restricted	40% to 50%
Dry matter	Depending on the fat in dry matter content, according to the table below			
	Fat in dry matter content (m/m)		**Corresponding minimum dry matter content** (m/m)	
	Equal to or above 40% but less than 60%		44%	
	Equal to or above 60%		54%	

Compositional modifications beyond the minima and maxima specified above for milkfat and dry matter are not considered to be in compliance with section 4.3.3 of the *General Standard for the Use of Dairy Terms* (CXS 206-1999).

4 FOOD ADDITIVES

Only those additives classes indicated as justified in the table below may be used for the product categories specified. Anticaking agents, colours and preservatives used in accordance with Tables 1 and 2 of the *General Standard for Food Additives* (CXS 192-1995) in food category 01.6.2.1 (Ripened cheese, includes rind) and only certain acidity regulators and anticaking agents in Table 3 are acceptable for use in foods conforming to this standard.

Additive functional class	Justified use	
	Cheese mass	Surface/rind treatment
Colours	×[a]	–
Bleaching agents	–	–
Acidity regulators	×	–
Stabilizers	–	–
Thickeners	–	–
Emulsifiers	–	–
Antioxidants	–	–

Additive functional class	Justified use	
	Cheese mass	Surface/rind treatment
Preservatives	×	×
Foaming agents	–	–
Anti–caking agents	–	×[(b)]

(a) Only to obtain the colour characteristics, as described in Section 2.

(b) For the surface of sliced, cut, shredded or grated cheese, only.

× The use of additives belonging to the class is technologically justified.

– The use of additives belonging to the class is not technologically justified.

5 CONTAMINANTS

The products covered by this Standard shall comply with the Maximum Levels for contaminants that are specified for the product in the *General Standard for Contaminants and Toxins in Food and Feed* (CXS 193-1995).

The milk used in the manufacture of the products covered by this Standard shall comply with the Maximum Levels for contaminants and toxins specified for milk by the *General Standard for Contaminants and Toxins in Food and Feed* (CXS 193-1995) and with the maximum residue limits for veterinary drug residues and pesticides established for milk by the CAC.

6 HYGIENE

It is recommended that the product covered by the provisions of this standard beprepared and handled in accordance with the appropriate sections of the *General Principles of Food Hygiene* (CXC 1-1969), the *Code of Hygienic Practice for Milk and Milk Products* (CXC 57-2004) and other relevant Codex texts such as Codes of Hygienic Practice and Codes of Practice. The products should comply with any microbiological criteria established in accordance with the *Principles and Guidelines for the Establishment and Application of Microbiological Criteria Related to Foods* (CXG 21-1997).

7 LABELLING

In addition to the provisions of the *General Standard for the Labelling of Prepackaged Foods* (CXS 1-1985) and the *General Standard for the Use of Dairy Terms* (CXS 206-1999), the following specific provisions apply.

7.1 Name of the food

The name Saint-Paulin may be applied in accordance with Section 4.1 of the *General Standard for the Labelling of Prepackaged Foods* (CXS 1-1985), provided that the product is in conformity with this Standard. Where customary in the country of retail sale, alternative spelling may be used.

The use of the name is an option that may be chosen only if the cheese complies with this standard. Where the name is not used for a cheese that complies with this standard, the naming provisions of the *General Standard for Cheese* (CXS 283-1978) apply.

The designation of products in which the fat content is above the reference range specified in Section 3.3 of this Standard shall be accompanied by an appropriate qualification describing the modification made or the fat content (expressed as fat in dry matter or as percentage by mass whichever is acceptable in the country of retail sale), either as part of the name or in a prominent position in the same field of vision. Suitable qualifiers are the appropriate characterizing terms specified in Section 7.2 of the *General Standard for Cheese* (CXS 283-1978) or a nutritional claim in accordance with the *Guidelines for Use of Nutrition and Health Claims* (CXG 23-1997)[1].

The designation may also be used for cut, sliced, shredded or grated products made from cheese which cheese is in conformity with this Standard.

7.2 Country of origin

The country of origin (which means the country of manufacture, not the country in which the name originated) shall be declared. When the product undergoes substantial transformation[2] in a second country, the country in which the transformation is performed shall be considered to be the country of origin for the purpose of labelling.

7.3 Declaration of milkfat content

The milk fat content shall be declared in a manner found acceptable in the country of retail sale either (i) as a percentage by mass, (ii) as a percentage of fat in dry matter, or (iii) in grams per serving as quantified in the label, provided that the number of servings is stated.

7.4 Date marking

Notwithstanding the provisions of Section 4.7.1 of the *General Standard for the Labelling of Prepackaged Foods* (CXS 1-1985), the date of manufacture may be declared instead of the minimum durability information, provided that the product is not intended to be purchased as such by the final consumer.

7.5 Labelling of non-retail containers

Information specified in Section 7 of this Standard and Sections 4.1 to 4.8 of the *General Standard for the Labelling of Prepackaged Foods* (CXS 1-1985) and, if necessary, storage instructions, shall be given either on the container or in accompanying documents, except that the name of the product, lot identification, and the name of the manufacturer or packer shall appear on the container, and in the absence of such a container, on the product itself. However, lot identification and the name and address may be replaced by an identification mark, provided that such mark is clearly identifiable with the accompanying documents.

8 METHODS OF SAMPLING AND ANALYSIS

For checking the compliance with this Standard, the methods of analysis and sampling contained in the *Recommended Methods of Analysis and Sampling* (CXS 234-1999) relevant to the provisions in this Standard, shall be used.

---

1 For the purpose of comparative nutritional claims, the minimum fat content of 40% fat in dry matter constitutes the reference.

2 For instance, repackaging, cutting, slicing, shredding and grating is not regarded as substantial transformation.

APPENDIX

ADDITIONAL INFORMATION

The additional information below does not affect the provisions in the preceding sections which are those that are essential to the product identity, the use of the name of the food and the safety of the food.

1 Appearance characteristics

1.1 Shape: Small flat cylinder with slightly convex sides. Other shapes are possible.

1.2 Dimensions and weights:

(a) Usual variant: Diameter approx. 20 cm; min. weight 1.3 kg.

(b) "Petit Saint-Paulin": Diameter 8–13 cm; min. weight 150 g.

(c) "Mini Saint-Paulin": Min. weight 20 g.

2 Method of manufacture

2.1 Fermentation procedure: Microbiologically derived acid development.

2.2 Other characteristics: The cheese is salted in brine.

3 Qualifiers

The designations "Petit Saint-Paulin" and "Mini Saint-Paulin" should be used when the cheese complies with the provisions for dimensions and weights (1.2).

菠萝伏洛干酪

STANDARD FOR PROVOLONE

CXS 272-1968

原为CODEX STAN C-15-1968。1968年通过。2007年修订。
2008年、2010年、2013年、2018年、2019年修订。

1 范围

本标准适用于符合本标准第2条所述即食或用于进一步加工的菠萝伏洛干酪。

2 说明

菠萝伏洛干酪是指符合《干酪通则》（CXS 283-1978）的成熟硬质或半硬质干酪。干酪外形呈圆柱形、梨形或其他形状，外观呈白色、银白色、淡黄色、橘黄色，为长条、平行的蛋白纤维质地，适合切割，如果时间较长也可以碾碎。一般无气孔，允许个别开口或裂口。干酪加工及销售时带或不带外皮均可[1]，也可以有外衣。

对于即食的菠萝伏洛干酪，调制味道和外观特征的催熟程序在10～20℃下通常需要1个月，取决于所需成熟度。在干酪具备和前述催熟程序同样的物理、生化和感官特性的情况下，可以使用其他催熟条件（包括添加催熟的酶）。如有合理技术或贸易需求，用于进一步加工的菠萝伏洛干酪无需具备同等成熟度。

菠萝伏洛干酪由"帕斯塔菲拉塔"加工工艺制作而成。即加热合适pH值的凝乳，揉捏或拉伸直到凝块平滑无结块。将凝乳趁热切割并入模成型，后冷却凝固。也可采用其他使成品干酪具有相同物理、生化和感官特性的加工工艺。

3 基本成分和质量要求

3.1 基本原料

牛乳、水牛乳或它们的混合乳，以及源于这些乳的制品。

3.2 其他配料

- 无害乳酸发酵剂和/或产香菌和其他无害微生物的培养剂；
- 凝乳酶或其他安全、适宜的凝固酶；
- 作为代盐制品的氯化钠和氯化钾；
- 饮用水；

---

[1] 并非在销售前就去掉外皮，相反干酪催熟后，以不产生外皮方式保存（无皮干酪）。在无皮干酪加工过程中，使用催熟膜。催熟膜也可以成为保护干酪的外衣。无皮干酪也可参见《干酪通则》（CXS 283-1978）附件。

- 提高催熟程序的安全、适宜的酶;
- 安全、适宜的加工助剂;
- 大米、玉米、马铃薯粉和淀粉:尽管有《干酪通则》(CXS 283-1978),但这些物质同样可作为抗结块剂用作处理切块、切片和切碎产品表面,只要添加量是良好操作规范所要求达到期待功能的必需量,同时考虑第4条规定的抗结块剂的使用。

3.3 成分

乳成分	最低含量（m/m）	最高含量（m/m）	参考含量（m/m）
干物质乳脂	45%	/	45%~50%
干物质	根据干物质脂肪含量,参见下表		
	干物质脂肪含量（m/m）	相应最低干物质含量（m/m）	
	≥45%,<50%	51%	
	≥50%,<60%	53%	
	≥60%	60%	

乳脂和干物质含量超过规定的最大值和最小值被视为不符合《乳品术语》(CXS 206-1999)第4.3.3条。

3.4 主要生产特点

主要发酵培养微生物应是瑞士乳杆菌、唾液链球菌亚种-嗜热链球菌、德氏乳酸杆菌亚种-保加利亚乳杆菌和干酪乳杆菌。

4 食品添加剂

只有下表中标注为合理使用的添加剂类别才可以用于指定产品类别。根据《食品添加剂通用标准》(CXS 192-1995)表1和表2在食品类别01.6.2.1(成熟干酪,包括外皮)中使用的抗结块剂、着色剂和防腐剂,以及表3中仅部分酸度调节剂和抗结块剂可用于符合本标准的食品。

添加剂功能类别	合理使用	
	干酪量	表面/外皮处理
着色剂	×[a]	-
漂白剂	-	-
酸度调节剂	×	-
稳定剂	-	-
增稠剂	-	-

添加剂功能类别	合理使用	
	干酪量	表面/外皮处理
乳化剂	–	–
抗氧化剂	–	–
防腐剂	×	×
发泡剂	–	–
抗结块剂	–	×(b)

（a）仅为获得第2条所述颜色特征。
（b）仅用于切片、切块、切碎或磨碎的乳酪表面。
× 使用该类别添加剂符合技术要求。
– 使用该类别添加剂不符合技术要求。

5 污染物

本标准所涉及的产品应符合《食品及饲料中污染物和毒素通用标准》（CXS 193-1995）规定的污染物最大限量。

本标准所涉及的产品在加工中使用的牛乳应符合《食品及饲料中污染物和毒素通用标准》（CXS 193-1995）规定的乳中污染物和毒素最大限量以及国际食品法典委员会设定的农药和兽药最大残留限量。

6 卫生要求

建议本标准所涉及的产品应遵循《食品卫生总则》（CXC 1-1969）、《乳及乳制品卫生操作规范》（CXC 57-2004）以及《卫生操作规范》和《生产操作规范》等其他相关法典文本。产品应符合《食品微生物标准制定与实施原则和准则》（CXG 21-1997）规定的所有微生物标准。

7 标识

除符合《预包装食品标识通用标准》（CXS 1-1985）和《乳品术语》（CXS 206-1999）外，还应符合下列具体规定。

7.1 产品名称

菠萝伏洛干酪名称应符合《预包装食品标识通用标准》（CXS 1-1985）第4.1条，只要产品符合该标准，产品销售国的习惯拼写也可被使用。

只要干酪符合本标准，就可选择使用此名称。对符合本标准但不使用此名称的干酪，应按《干酪通则》（CXS 283-1978）命名规定来命名。

对于脂肪含量超过本标准第3.3条规定的参考含量范围，但高于绝对最低含量的产品，在命名时须附上适当修饰语，作为名称的一部分或处于与名称同一视野内的显眼位置，说明所做调整或脂肪含量（以干物质中脂肪含量或质量百分比表示，视产品销售国的接受情况）。适当修饰语是指《干酪通则》（CXS 283-1978）第7.2条所规定的特征描述用语或是符合《营养和保健声明使用准则》（CXG 23-1997）[1]的营养声明。

该名称也可用于符合本标准规定的干酪做成的切块、切片、切碎或磨碎产品。

7.2 原产国

应当标记原产国（即生产国，而非名称起源国）。当产品在第二个国家有实质性改造[2]，则以标识目的进行改造的国家应视为原产国。

7.3 乳脂含量说明

乳脂含量应以销售国可接受的方式声明：（i）质量百分比形式；（ii）干物质中脂肪百分比形式；（iii）如果产品标识上标明了份数，也可用每份中乳脂重量（g）表示。

7.4 日期标识

尽管《预包装食品标识通用标准》（CXS 1-1985）第4.7.1条规定，只要产品并非直接出售给最终消费者，仍应标明生产日期而非最低保存时间信息。

7.5 非零售包装标识

在包装容器上除要标明产品名称、批次、生产厂家或包装商的名称和地址外，在包装容器上或附随说明书中，也应按本标准第7条和《预包装食品标识通用标准》（CXS 1-1985）第4.1~4.8条所要求的信息以及储存说明（在必要的地方）加以陈述。如果批次、生产厂家或包装商的名称和地址可以在附随的文件标明，则可以用一个识别标识来代替。

8 抽样和分析方法

为检查产品是否符合本标准，应采用《分析和抽样推荐性方法》（CXS 234-1999）中涉及本标准规定的分析和抽样方法。

---

1 为比较营养声明，干物质中脂肪含量最低参考值为45%。
2 例如，重新包装、切块、切片、切碎和磨碎不构成实质性改变。

附录

附加信息

以下补充信息对前文各条款的规定不构成影响,前文规定内容对产品标识、食品名称的使用以及食品安全性至关重要。

1 外观特征

1.1 典型形状:圆柱状(Salame)、梨形(Mandarino)、梨圆柱状(Gigantino)和长颈瓶状(Fiaschetta)。

1.2 典型包装:干酪通常用绳子包裹起来。

STANDARD FOR PROVOLONE

CXS 272-1968

Formerly CODEX STAN C-15-1968. Adopted in 1968. Revised in 2007.

Amended in 2008, 2010, 2013, 2018, 2019.

1 SCOPE

This Standard applies to Provolone intended for direct consumption or for further processing in conformity with the description in Section 2 of this Standard.

2 DESCRIPTION

Provolone is a ripened firm/semi-hard cheese in conformity with the *General Standard for Cheese* (CXS 283-1978). The body has a near white or ivory through to light yellow or yellow colour and a fibrous texture with long stranded parallel-orientated protein fibres. It is suitable for cutting and, when aged, for grating as well. Gas holes are generally absent, but few openings and splits are acceptable. The shape is mainly cylindrical or pear-shaped, but other shapes are possible. The cheese is manufactured and sold with or without[1] a rind, which may be coated.

For Provolone ready for consumption, the ripening procedure to develop flavour and body characteristics is normally from 1 month at 10–20 °C depending on the extent of maturity required. Alternative ripening conditions (including the addition of ripening enhancing enzymes) may be used, provided the cheese exhibits similar physical, biochemical and sensory properties as those achieved by the previously stated ripening procedure. Provolone intended for further processing and Provolone of low weights (< 2 kg) need not exhibit the same degree of ripening when justified through technical and/or trade needs.

Provolone is made by "pasta filata" processing which consists of heating curd of a suitable pH value, kneading and stretching until the curd is smooth and free from lumps. Still warm, the curd is cut and moulded, then firmed by cooling in chilled water or brine. Other processing techniques, which give end products with the same physical, chemical and organoleptic characteristics are allowed.

3 ESSENTIAL COMPOSITION AND QUALITY FACTORS

3.1 Raw materials

Cows' milk or buffaloes' milk, or their mixtures, and products obtained from these milks.

3.2 Permitted ingredients

– Starter cultures of harmless lactic acid and/or flavour producing bacteria and cultures of other harmless micro-organisms;

---

[1] This is not to mean that the rind has been removed before sale, instead the cheese has been ripened and/or kept in such a way that no rind is developed (a "rindless" cheese). Ripening film is used in the manufacture of rindless cheese. Ripening film may also constitute the coating that protects the cheese. For rindless cheese see also the Appendix to the *General Standard for Cheese* (CXS 283-1978).

- Rennet or other safe and suitable coagulating enzymes;
- Sodium chloride and potassium chloride as salt substitute;
- Safe and suitable enzymes to enhance the ripening process;
- Safe and suitable processing aids;
- Potable water;
- Rice, corn and potato flours and starches: Notwithstanding the provisions in the *General Standard for Cheese* (CXS 283-1978), these substances can be used in the same function as anti-caking agents for treatment of the surface of cut, sliced, and shredded products only, provided they are added only in amounts functionally necessary as governed by Good Manufacturing Practice, taking into account any use of the anti-caking agents listed in Section 4.

3.3 Composition

Milk constituent	Minimum content (m/m)	Maximum content (m/m)	Reference level (m/m)
Milk fat in dry matter	45%	Not restricted	45% to 50%
Dry matter	Depending on the fat in dry matter content, according to the table below		
	Fat in dry matter content (m/m)		**Corresponding minimum dry matter content** (m/m)
	Equal to or above 45% but less than 50%		51%
	Equal to or above 50% but less than 60%		53%
	Equal to or above 60%		60%

Compositional modifications beyond the minima and maxima specified above for milk fat and dry matter are not considered to be in compliance with Section 4.3.3 of the *General Standard for the Use of Dairy Terms* (CXS 206-1999).

3.4 Essential manufacturing characteristics

The principal starter culture micro-organisms shall be *Lactobacillus helveticus*, *Streptococcus salivarius* subsp. *thermophilus*, *Lactobacillusdelbrueckii* subsp. *bulgaricus* and *Lactobacillus casei*.

4 FOOD ADDITIVES

Only those additives classes indicated as justified in the table below may be used for the product categories specified. Anticaking agents, colours and preservatives used in accordance with Tables 1 and 2 of the *General Standard for Food Additives* (CXS 192-1995) in food category 01.6.2.1 (Ripened cheese, includes rind) and only certain acidity regulators, anticaking agents and colours in Table 3 are acceptable for use in foods conforming to this standard.

Additive functional class	Justified use	
	Cheese mass	Surface/rind treatment
Colours	x[a]	–
Bleaching agents	–	–

Additive functional class	Justified use	
	Cheese mass	Surface/rind treatment
Acidity regulators	×	–
Stabilizers	–	–
Thickeners	–	–
Emulsifiers	–	–
Antioxidants	–	–
Preservatives	×	×
Foaming agents	–	–
Anti–caking agents	–	×[b]

(a) Only to obtain the colour characteristics, as described in Section 2.
(b) For the surface of sliced, cut, shredded or grated cheese, only.
× The use of additives belonging to the class is technologically justified.
– The use of additives belonging to the class is not technologically justified.

5 CONTAMINANTS

The products covered by this Standard shall comply with the Maximum Levels for contaminants that are specified for the product in the *General Standard for Contaminants and Toxins in Food and Feed* (CXS 193-1995).

The milk used in the manufacture of the products covered by this Standard shall comply with the Maximum Levels for contaminants and toxins specified for milk by the *General Standard for Contaminants and Toxins in Food and Feed* (CXS 193-1995) and with the maximum residue limits for veterinary drug residues and pesticides established for milk by the CAC.

6 HYGIENE

It is recommended that the product covered by the provisions of this standard be prepared and handled in accordance with the appropriate sections of the *General Principles of Food Hygiene* (CXC 1-1969), the *Code of Hygienic Practice for Milk and Milk Products* (CXC 57-2004) and other relevant Codex texts such as Codes of Hygienic Practice and Codes of Practice. The products should comply with any microbiological criteria established in accordance with the *Principles and Guidelines for the Establishment and Application of Microbiological Criteria Related to Foods* (CXG 21-1997).

7 LABELLING

In addition to the provisions of the *General Standard for the Labelling of Prepackaged Foods* (CXS 1-1985) and the *General Standard for the Use of Dairy Terms* (CXS 206-1999), the following specific provisions apply.

7.1 Name of the food

The name Provolone may be applied in accordance with Section 4.1 of the *General Standard for the Labelling of Prepackaged Foods* (CXS 1-1985), provided that the product is in conformity with this Standard. Where customary in the country of retail sale, alternative spelling may be used.

The use of the name is an option that may be chosen only if the cheese complies with this standard. Where the name is not used for a cheese that complies with this standard, the naming provisions of the *General Standard for Cheese* (CXS 283-1978) apply.

The designation of products in which the fat content is above the reference range specified in Section 3.3 of this Standard shall be accompanied by an appropriate qualification describing the modification made or the fat content (expressed as fat in dry matter or as percentage by mass whichever is acceptable in the country of retail sale), either as part of the name or in a prominent position in the same field of vision. Suitable qualifiers are the appropriate characterizing terms specified in Section 7.2 of the *General Standard for Cheese* (CXS 283-1978) or a nutritional claim in accordance with the *Guidelines for Use of Nutrition and Health Claims* (CXG 23-1997).[1]

The designation may also be used for cut, sliced, shredded or grated products made from cheese which cheese is in conformity with this Standard.

7.2 Country of origin

The country of origin (which means the country of manufacture, not the country in which the name originated) shall be declared. When the product undergoes substantial transformation[2] in a second country, the country in which the transformation is performed shall be considered to be the country of origin for the purpose of labelling.

7.3 Declaration of milk fat content

The milk fat content shall be declared in a manner found acceptable in the country of retail sale either (i) as a percentage by mass, (ii) as a percentage of fat in dry matter, or (iii) in grams per serving as quantified in the label, provided that the number of servings is stated.

7.4 Date marking

Notwithstanding the provisions of Section 4.7.1 of the *General Standard for the Labelling of Prepackaged Foods* (CXS 1-1985), the date of manufacture may be declared instead of the minimum durability information, provided that the product is not intended to be purchased as such by the final consumer.

7.5 Labelling of non-retail containers

Information specified in Section 7 of this Standard and Sections 4.1 to 4.8 of the *General Standard for the Labelling of Prepackaged Foods* (CXS 1-1985) and, if necessary, storage instructions, shall be given either on the container or in accompanying documents, except that the name of the product, lot identification, and the name of the manufacturer or packer shall appear on the container, and in the absence of such a container, on the product itself. However, lot identification and the name and address may be replaced by an identification mark, provided that such mark is clearly identifiable with the accompanying documents.

8 METHODS OF SAMPLING AND ANALYSIS

For checking the compliance with this standard, the methods of analysis and sampling contained in the *Recommended Methods of Analysis and Sampling* (CXS 234-1999) relevant to the provisions in this standard, shall be used.

---

1 For the purpose of comparative nutritional claims, the average minimum fat content of 45% fat in dry matter constitutes the reference.

2 For instance, repackaging, cutting, slicing, shredding and grating is not regarded as substantial transformation.

APPENDIX

ADDITIONAL INFORMATION

The additional information below does not affect the provisions in the preceding sections which are those that are essential to the product identity, the use of the name of the food and the safety of the food.

1   **Appearance characteristics**

1.1   Typical shapes: Cylindrical (Salame), pear-shaped (Mandarino), pear-shaped cylinder (Gigantino) and flask (Fiaschetta).

1.2   Typical packing: The cheese is typically encased in ropes.

农家干酪

STANDARD FOR COTTAGE CHEESE

CXS 273-1968

前为CODEX STAN C-16-1968。1968年通过。

2007年、2010年、2018年修订。2014年、2016年修正。

1　范围

本标准适用于符合本标准第2条所述即食或用于进一步加工的农家干酪。

2　说明

农家干酪是指符合《干酪通则》（CXS 283-1978）和《未成熟干酪（包括新鲜干酪）标准》（CXS 221-2001）的软质、可涂抹的、未成熟的无外皮[1]干酪。酪体近乎白色，颗粒状质地，由离散的各个软质凝块颗粒组成，颗粒大小相对一致，视理想的凝块大小而定，有3～12 mm不等，可能覆有乳状混合物。

3　基本成分和质量要求

3.1　基本原料

牛乳、水牛乳或它们的混合乳，以及源于这些乳的制品。

3.2　其他配料

– 无害乳酸发酵剂和/或产香菌和其他无害微生物的培养剂；

– 凝乳酶或其他安全、适宜的凝固酶；

– 明胶和淀粉：这些物质可以作为相同功能稳定剂使用，但添加剂量仅以发挥必要的功能为限，要以考虑使用第4条所列稳定剂/增稠剂的良好生产规范为准；

– 作为代盐制品的氯化钠和氯化钾；

– 饮用水；

– 安全、适宜的加工助剂。

3.3　成分

乳成分	最低含量（m/m）	最高含量（m/m）	参考含量（m/m）
乳脂	0%	/	4%～5%
脱脂干物质	18%	以脱脂水分为限	

---

[1]　这种干酪在保存中不会结成外皮（"无外皮"干酪）。

乳脂和干物质含量超过规定的最大值和最小值被视为不符合《乳品术语》（CXS 206-1999）第4.3.3条。

4　食品添加剂

只允许在规定范围内使用下表列出的食品添加剂。

添加剂功能类别	合理使用	
	干酪量[b]	表面/外皮处理
着色剂	–	–
漂白剂	–	–
酸度调节剂	×	–
稳定剂	×[a]	–
增稠剂	–	–
乳化剂	–	–
抗氧化剂	–	–
防腐剂	×	–
发泡剂	–	–
抗结块剂	–	–

（a）可按乳制品定义使用包括改性淀粉在内的稳定剂，但仅以发挥必要的功能为限，用于加热并考虑使用第3.2条所述明胶和淀粉。

（b）干酪质量包括乳状混合物。

× 使用该类别添加剂符合技术要求。

– 使用该类别添加剂不符合技术要求。

INS编号	添加剂名称	最大限量
防腐剂		
200	山梨酸	1 000 mg/kg 单用或混用，以山梨酸计
202	山梨酸钾	
203	山梨酸钙	
234	乳酸链球菌素	12.5 mg/kg
280	丙酸	根据GMP限量使用
281	丙酸钠	
282	丙酸钙	
283	丙酸钾	

INS编号	添加剂名称	最大限量
酸度调节剂		
170（i）	碳酸钙	根据GMP限量使用
260	冰醋酸	
261（i）	醋酸钾	
261（ii）	双乙酸钾	
262（i）	乙酸钠	
263	乙酸钙	
270	乳酸（L-，D-和DL-）	
296	DL-苹果酸	
325	乳酸钠	
326	乳酸钾	
327	乳酸钙	
330	柠檬酸	
338	磷酸	880 mg/kg，以磷计
350（i）	DL-苹果酸氢钠	根据GMP限量使用
350（ii）	DL-苹果酸钠	
352（ii）	D，L-苹果酸钙	
500（i）	碳酸钠	
500（ii）	碳酸氢钠	
500（iii）	倍半碳酸钠	
501（i）	碳酸钾	
501（ii）	碳酸氢钾	
504（i）	碳酸镁	
504（ii）	碳酸氢镁	
507	盐酸	
575	葡萄糖酸-δ-内酯	
577	葡萄糖酸钾	
578	葡萄糖酸钙	

INS编号	添加剂名称	最大限量
稳定剂		
331（i）	柠檬酸二氢钠	根据GMP限量使用
332（i）	柠檬酸二氢钾	
333	柠檬酸钙	
339（i）	磷酸二氢钠	1 300 mg/kg 单用或混用，以磷计
339（ii）	磷酸氢二钠	
339（iii）	磷酸三钠	
340（i）	磷酸二氢钾	
340（ii）	磷酸氢二钾	
340（iii）	磷酸三钾	
341（i）	磷酸二氢钙	
341（ii）	磷酸氢二钙	
341（iii）	磷酸三钙	
342（i）	磷酸二氢铵	
342（ii）	磷酸氢二铵	
343（ii）	磷酸氢二镁	
343（iii）	磷酸三镁	
450（i）	磷酸二钠	
450（iii）	焦磷酸四钠	
450（v）	二磷酸四钾	
450（vi）	磷酸二钙	
451（i）	三磷酸五钠	
451（ii）	三磷酸五钾	
452（i）	六偏磷酸钠	
452（ii）	聚磷酸钾	
452（iv）	聚磷酸钙	
452（v）	聚磷酸铵	

INS编号	添加剂名称	最大限量
400	海藻酸	根据GMP限量使用
401	海藻酸钠	
402	海藻酸钾	
403	海藻酸铵	
404	海藻酸钙	
405	海藻酸丙二醇酯	5 000 mg/kg
406	琼脂	根据GMP限量使用
407	卡拉胶	
407a	加工琼芝属海藻胶（PES）	
410	槐豆胶（又名刺槐豆胶）	
412	瓜尔胶	
413	黄蓍胶	
415	黄原胶	
416	刺梧桐胶	
417	刺云实胶	
440	果胶	
466	羧甲基纤维素钠（纤维素胶）	
1400	糊精，焙炒淀粉	
1401	酸处理淀粉	
1402	碱处理淀粉	
1403	漂白淀粉	
1404	氧化淀粉	
1405	酶处理淀粉	
1410	单淀粉磷酸酯	
1412	磷酸酯双淀粉	
1413	磷酸化二淀粉磷酸酯	
1414	乙酰化二淀粉磷酸酯	
1420	醋酸酯淀粉	
1422	乙酰化双淀粉己二酸酯	

INS编号	添加剂名称	最大限量
1440	羟丙基淀粉	根据GMP限量使用
1442	羟丙基二淀粉磷酸酯	

5 污染物

本标准所涉及的产品应符合《食品及饲料中污染物和毒素通用标准》（CXS 193-1995）规定的污染物最大限量。

本标准所涉及的产品在加工中使用的牛乳应符合《食品及饲料中污染物和毒素通用标准》（CXS 193-1995）规定的乳中污染物和毒素最大限量以及国际食品法典委员会设定的农药和兽药最大残留限量。

6 卫生要求

建议本标准所涉及的产品应遵循《食品卫生总则》（CXC 1-1969）、《乳及乳制品卫生操作规范》（CXC 57-2004）以及《卫生操作规范》和《生产操作规范》等其他相关法典文本。产品应符合《食品微生物标准制定与实施原则和准则》（CXG 21-1997）规定的所有微生物标准。

7 标识

除符合《预包装食品标识通用标准》（CXS 1-1985）和《乳品术语》（CXS 206-1999）外，还应符合下列具体规定。

7.1 产品名称

农家干酪应符合《预包装食品标识通用标准》（CXS 1-1985）第4.1条，只要产品符合该标准，产品销售国的习惯拼写也可被使用。

只要干酪符合本标准，就可选择使用此名称。对符合本标准但不使用此名称的干酪，应按《干酪通则》（CXS 283-1978）命名规定来命名。

对于脂肪含量超过本标准第3.3条规定的参考含量范围，但高于绝对最低含量的产品，在命名时须附上适当修饰语，作为名称的一部分或处于与名称同一视野内的显眼位置，说明所做的调整或脂肪含量（以干物质中脂肪含量或质量百分比表示，视产品销售国的接受情况）。适当修饰语包括《营养和保健声明使用准则》（CXG 23-1997）[1]提出的营养声明。此外，可在食品名称上加注说明产品性质或特点的相应特征界定术语。此类术语包括"干凝乳"或"奶油"。

7.2 原产国

应当标记原产国（即生产国，而非名称起源国）。当产品在第二个国家有实质性改

---

[1] 为比较营养声明，干物质中脂肪含量最低参考值为4%。

造[1]，则以标识目的进行改造的国家应视为原产国。

7.3 乳脂含量说明

乳脂含量应以销售国可接受的方式声明：（i）质量百分比形式；（ii）干物质中脂肪百分比形式；（iii）如果产品标识上标明了份数，也可用每份中乳脂重量（g）表示。

7.4 非零售包装标识

在包装容器上除要标明产品名称、批次、生产厂家或包装商的名称和地址外，在包装容器上或附随说明书中，也应按本标准第7条和《预包装食品标识通用标准》（CXS 1-1985）第4.1~4.8条所要求的信息以及储存说明（在必要的地方）加以陈述。如果批次、生产厂家或包装商的名称和地址可以在附随的文件标明，则可以用一个识别标识来代替。

8 抽样和分析方法

为检查产品是否符合本标准，应采用《分析和抽样推荐性方法》（CXS 234-1999）中涉及本标准规定的分析和抽样方法。

---

[1] 例如，重新包装、切块、切片、切碎和磨碎不构成实质性改变。

STANDARD FOR COTTAGE CHEESE

CXS 273-1968

Formerly CODEX STAN C-16-1968. Adopted in 1968.
Revised in 2007, 2010, 2018. Amended in 2014, 2016.

1 SCOPE

This Standard applies to Cottage Cheese intended for direct consumption or for further processing in conformity with the description in Section 2 of this Standard.

2 DESCRIPTION

Cottage Cheese is a soft, rindless[1], unripened cheese in conformity with the *General Standard for Cheese* (CXS 283-1978) and the *Group Standard for Unripened Cheese Including Fresh Cheese* (CXS 221-2001). The body has a near white colour and a granular texture consisting of discrete individual soft curd granules of relatively uniform size, from approximately 3–12 mm depending on whether small or large type of curd is desired, and possibly covered with a creamy mixture.

3 ESSENTIAL COMPOSITION AND QUALITY FACTORS

3.1 Raw materials

Cows' milk or buffaloes' milk, or their mixtures, and products obtained from these milks.

3.2 Permitted ingredients

- Starter cultures of harmless lactic acid and/or flavour producing bacteria and cultures of other harmless micro-organisms;
- Rennet or other safe and suitable coagulating enzymes;
- Gelatin and starches: These substances can be used in the same function as stabilizers, provided they are added only in amounts functionally necessary as governed by Good Manufacturing Practice taking into account any use of the stabilizers/thickeners listed in Section 4;
- Sodium chloride and potassium chloride as a salt substitute;
- Potable water;
- Safe and suitable processing aids.

3.3 Composition

Milk constituent	Minimum content (m/m)	Maximum content (m/m)	Reference level (m/m)
Milkfat	0%	Not restricted	4%–5%
Fat free dry matter	18%	Restricted by the MFFB	

---

[1] The cheese has been kept in such a way that no rind is developed (a "rindless" cheese).

Compositional modifications beyond the minimum and maximum specified above for fat free dry matter are not considered to be in compliance with section 4.3.3 of the *General Standard for the Use of Dairy Terms* (CXS 206-1999).

4 FOOD ADDITIVES

Only those additives classes indicated as justified in the table below may be used for the product categories specified. Within each additive class, and where permitted according to the table, only those food additives listed below may be used and only within the functions and limits specified.

Additive functional class	Justified use	
	Cheese mass[b]	Surface/rind treatment
Colours	–	–
Bleaching agents	–	–
Acidity regulators	×	–
Stabilizers	×[a]	–
Thickeners	–	–
Emulsifiers	–	–
Antioxidants	–	–
Preservatives	×	–
Foaming agents	–	–
Anti–caking agents	–	–

(a) Stabilizers including modified starches may be used in compliance with the definition of milk products and only to the extent they are functionally necessary, taking into account any use of gelatine and starches as provided for in Section 3.2.

(b) Cheese mass includes creaming mixture.

× The use of additives belonging to the class is technologically justified.

– The use of additives belonging to the class is not technologically justified.

INS no.	Name of additive	Maximum level
Preservatives		
200	Sorbic acid	1 000 mg/kg, singly or in combinations sorbic acid
202	Potassium sorbate	
203	Calcium sorbate	
234	Nisin	12.5 mg/kg
280	Propionic acid	Limited by GMP
281	Sodium propionate	

INS no.	Name of additive	Maximum level
282	Calcium propionate	Limited by GMP
283	Potassium propionate	
Acidity regulators		
170(i)	Calcium carbonate	Limited by GMP
260	Acetic acid, glacial	
261(i)	Potassium acetate	
261(ii)	Potassium diacetate	
262(i)	Sodium acetate	
263	Calcium acetate	
270	Lactic acid, *L-, D-* and *DL-*	
296	Malic acid, *DL-*	
325	Sodium lactate	
326	Potassium lactate	
327	Calcium lactate	
330	Citric acid	
338	Phosphoric acid	880 mg/kg as phosphorous
350(i)	Sodium hydrogen *DL*-malate	Limited by GMP
350(ii)	Sodium *DL*-malate	
352(ii)	Calcium malate, *D, L-*	
500(i)	Sodium carbonate	
500(ii)	Sodium hydrogen carbonate	
500(iii)	Sodium sesquicarbonate	
501(i)	Potassium carbonate	
501(ii)	Potassium hydrogen carbonate	
504(i)	Magnesium carbonate	
504(ii)	Magnesium hydrogen carbonate	
507	Hydrochloric acid	
575	Glucono-*delta*-lactone	
577	Potassium gluconate	
578	Calcium gluconate	

INS no.	Name of additive	Maximum level
Stabilizers		
331(i)	Sodium dihydrogen citrate	Limited by GMP
332(i)	Potassium dihydrogen citrate	
333	Calcium citrates	
339(i)	Sodium dihydrogen phosphate	1 300 mg/kg, singly or in combination, expressed as phosphorus
339(ii)	Disodium hydrogen phosphate	
339(iii)	Trisodium phosphate	
340(i)	Potassium dihydrogen phosphate	
340(ii)	Dipotassium hydrogen phosphate	
340(iii)	Tripotassium phosphate	
341(i)	Calcium dihydrogen phosphate	
341(ii)	Calcium hydrogen phosphate	
341(iii)	Tricalcium phosphate	
342(i)	Ammonium dihydrogen phosphate	
342(ii)	Ammonium hydrogen phosphate	
343(ii)	Magnesium hydrogen phosphate	
343(iii)	Trimagnesium phosphate	
450(i)	Disodium diphosphate	
450(iii)	Tetrasodium diphosphate	
450(v)	Tetrapotassium diphosphate	
450(vi)	Dicalcium diphosphate	
451(i)	Pentasodium triphosphate	
451(ii)	Pentapotassium triphosphate	
452(i)	Sodium polyphosphate	
452(ii)	Potassium polyphosphate	
452(iv)	Calcium polyphosphate	
452(v)	Ammonium polyphosphate	
400	Alginic acid	Limited by GMP
401	Sodium alginate	

INS no.	Name of additive	Maximum level
402	Potassium alginate	Limited by GMP
403	Ammonium alginate	
404	Calcium alginate	
405	Propylene glycol alginate	5 000 mg/kg
406	Agar	Limited by GMP
407	Carrageenan	
407a	Processed euchema seaweed (PES)	
410	Carob bean gum	
412	Guar gum	
413	Tragacanth gum	
415	Xanthan gum	
416	Karaya gum	
417	Tara gum	
440	Pectins	
466	Sodium carboxymethyl cellulose (Cellulose gum)	
1400	Dextrins, roasted starch	
1401	Acid-treated starch	
1402	Alkaline-treated starch	
1403	Bleached starch	
1404	Oxidized starch	
1405	Starches, enzyme-treated	
1410	Monostarch phosphate	
1412	Distarch phosphate	
1413	Phosphateddistarch phosphate	
1414	Acetylated distarch phosphate	
1420	Starch acetate	
1422	Acetylated distarchadipate	
1440	Hydroxypropyl starch	
1442	Hydroxypropyldistarch phosphate	

5　**CONTAMINANTS**

The products covered by this Standard shall comply with the maximum levels for contaminants that are specified for the product in the *General Standard for Contaminants and Toxins in Food and Feed* (CXS 193-1995).

The milk used in the manufacture of the products covered by this Standard shall comply with the maximum levels for contaminants and toxins specified for milk by the *General Standard for Contaminants and Toxins in Food and Feed* (CXS 193-1995) and with the maximum residue limits for veterinary drug residues and pesticides established for milk by the CAC.

6　**HYGIENE**

It is recommended that the product covered by the provisions of this standard be prepared and handled in accordance with the appropriate sections of the *General Principles of Food Hygiene* (CXC 1-1969), the *Code of Hygienic Practice for Milk and Milk Products* (CXC 57-2004) and other relevant Codex texts such as Codes of Hygienic Practice and Codes of Practice.

The products should comply with any microbiological criteria established in accordance with the *Principles and Guidelines for the Establishment and Application of Microbiological Criteria Related to Foods* (CXG 21-1997).

7　**LABELLING**

In addition to the provisions of the *General Standard for the Labelling of Prepackaged Foods* (CXS 1-1985) and the *General Standard for the Use of Dairy Terms* (CXS 206-1999), the following specific provisions apply.

7.1　**Name of the food**

The name Cottage Cheese may be applied in accordance with Section 4.1 of the *General Standard for the Labelling of Prepackaged Foods* (CXS 1-1985), provided that the product is in conformity with this Standard. Where customary in the country of retail sale, alternative spelling may be used. The name may be translated into other languages so that the consumer in the country of retail sale will not be mislead.

The use of the name is an option that may be chosen only if the cheese complies with this Standard. Where the name is not used for a cheese that complies with this standard, the naming provisions of the *General Standard for Cheese* (CXS 283-1978) apply.

The designation of products in which the fat content is below or above the reference range specified in Section 3.3 of this Standard shall be accompanied by an appropriate qualification describing the modification made or the fat content (expressed as fat in dry matter or as percentage by mass whichever is acceptable in the country of retail sale), either as part of the name or in a prominent position in the same field of vision. Suitable qualifiers include nutritional claims in accordance with the *Guidelines for Use of Nutrition and Health Claims*[1] (CXG 23-1997). In addition the appropriate characterizing terms describing the nature or style of the product may accompany the name of the food. Such terms include "dry curd" or "creamed".

1　For the purpose of comparative nutritional claims, the fat content of 4% constitutes the reference.

7.2 Country of origin

The country of origin (which means the country of manufacture, not the country in which the name originated) shall be declared. When the product undergoes substantial transformation[1] in a second country, the country in which the transformation is performed shall be considered to be the country of origin for the purpose of labelling.

7.3 Declaration of milkfat content

The milk fat content shall be declared in a manner found acceptable in the country of retail sale, either (i) as a percentage by mass, (ii) as a percentage of fat in dry matter, or (iii) in grams per serving as quantified in the label, provided that the number of servings is stated.

7.4 Labelling of non-retail containers

Information specified in Section 7 of this Standard and Sections 4.1 to 4.8 of the *General Standard for the Labelling of Prepackaged Foods* (CXS 1-1985) and, if necessary, storage instructions, shall be given either on the container or in accompanying documents, except that the name of the product, lot identification, and the name of the manufacturer or packer shall appear on the container, and in the absence of such a container, on the product itself. However, lot identification and the name and address may be replaced by an identification mark, provided that such mark is clearly identifiable with the accompanying documents.

8 METHODS OF SAMPLING AND ANALYSIS

For checking the compliance with this Standard, the methods of analysis and sampling contained in the *Recommended Methods of Analysis and Sampling* (CXS 234-1999) relevant to the provisions in this Standard, shall be used.

---

[1] For instance, repackaging, cutting, slicing, shredding and grating is not regarded as substantial transformation.

库洛米耶尔干酪
STANDARD FOR COULOMMIERS
CXS 274-1969

前为CODEX STAN C-18-1969。1969年通过。2007年修订。

2008年、2010年、2018年、2019年修正。

1 范围

本标准适用于符合本标准第2条所述即食或用于进一步加工的库洛米耶尔干酪。

2 说明

库洛米耶尔干酪是指符合《干酪通则》（CXS 283-1978）的软质且表面成熟的干酪（多为霉菌成熟干酪），呈扁平圆柱形或扇形。干酪外呈近白色，内呈浅黄色，其质地绵软（用拇指摁压）、不松脆，从外到里均已经成熟。通常不见气孔，但允许少量裂口。理想的外皮质地绵软，通体覆以白色霉菌，但可布有红色、褐色或橙色色斑。霉菌生长前后，整块干酪可以切成或塑成扇形。

对于即食的库洛米耶尔干酪，调制味道和外观特征的催熟程序通常需要10天，温度需为10~16 ℃，取决于所需成熟度。在干酪具备和前述催熟程序同样的物理、生化和感官特性的情况下，也可采用其他熟化条件（包括添加加速催熟的酶）。如有合理技术或贸易合理需求，用于进一步加工的库洛米耶尔干酪无需具备同等成熟度。

3 基本成分和质量要求

3.1 基本原料

牛乳、水牛乳或它们的混合乳，以及源于这些乳的制品。

3.2 其他配料

- 无害乳酸发酵剂和/或产香菌和其他无害微生物（包括白地霉、亚麻短杆菌和酵母菌）的培养物；
- 凝乳酶或其他安全、适用的凝固酶；
- 作为代盐制品的氯化钠和氯化钾；
- 饮用水；
- 安全、适宜的加工助剂；
- 促进成熟的安全、适宜的酶；

— 大米、玉米、马铃薯粉和淀粉：尽管有《干酪通则》（CXS 283-1978），但这些物质同样可作为抗结块剂处理切块、切片和切碎产品表面，只要添加量是良好操作规范所要求达到期待功能的必需量，同时考虑第4条规定的抗结块剂的使用。

3.3 成分

乳成分	最低含量（m/m）	最高含量（m/m）	参考含量（m/m）
干物质乳脂	40%	/	40%～50%
干物质	根据干物质脂肪含量，参见下表		
	干物质脂肪含量（m/m）	相应最小干物质含量（m/m）	
	≥40%，<50%	42%	
	≥50%，<60%	46%	
	≥60%	52%	

乳脂和干物质含量超过规定的最大值和最小值被视为不符合《乳品术语》（CXS 206-1999）第4.3.3条。

3.4 基本尺寸和形状

最大高度：大约5 cm；

重量：整个平柱干酪：最小300 g。

3.5 基本熟化过程

从外到里的结皮和成熟（蛋白酶解）主要是卡地干酪青霉和/或卡门柏青霉和白酪青霉作用。

4 食品添加剂

只有下表所示可以合理使用的添加剂类别才可用于指定产品类别。根据《食品添加剂通用标准》（CXS 192-1995）表1和表2在食品类别01.6.2.1（成熟干酪，包括外皮）中使用的着色剂，以及表3中仅部分酸度调节剂可用于符合本标准的食品。

添加剂功能类别	合理使用	
	干酪量	表面/外皮处理
着色剂	×[a]	—
漂白剂	—	—
酸度调节剂	×	—

添加剂功能类别	合理使用	
	干酪量	表面/外皮处理
稳定剂	–	–
增稠剂	–	–
乳化剂	–	–
抗氧化剂	–	–
防腐剂	–	–
发泡剂	–	–
抗结块剂	–	–

（a）仅为获得第2条所述颜色特征。

× 使用该类别添加剂符合技术要求。

– 使用该类别添加剂不符合技术要求。

5　污染物

本标准所涉及的产品应符合《食品及饲料中污染物和毒素通用标准》（CXS 193-1995）规定的污染物最大限量。

本标准所涉及的产品在加工中使用的牛乳应符合《食品及饲料中污染物和毒素通用标准》（CXS 193-1995）规定的乳中污染物和毒素最大限量以及国际食品法典委员会设定的农药和兽药最大残留限量。

6　卫生要求

建议本标准所涉及的产品应遵循《食品卫生总则》（CXC 1-1969）、《乳及乳制品卫生操作规范》（CXC 57-2004）以及《卫生操作规范》和《生产操作规范》等其他相关法典文本。产品应符合《食品微生物标准制定与实施原则和准则》（CXG 21-1997）规定的所有微生物标准。

7　标识

除符合《预包装食品标识通用标准》（CXS 1-1985）和《乳品术语》（CXS 206-1999）外，还应符合下列具体规定。

7.1　产品名称

库洛米耶尔干酪名称应符合《预包装食品标识通用标准》（CXS 1-1985）第4.1条，只要产品符合该标准，产品销售国的习惯拼写也可被使用。

只要干酪符合本标准，就可选择使用此名称。对符合本标准但不使用此名称的干酪，应按《干酪通则》（CXS 283-1978）命名规定来命名。

对于脂肪含量超过本标准第3.3条规定的参考含量范围，但高于绝对最低含量的产品，在命名时须附上适当修饰语，作为名称的一部分或处于与名称同一视野内的显眼位置，说明所做调整或脂肪含量（以干物质中脂肪含量或质量百分比表示，视产品销售国的接受情况）。适当修饰语是指《干酪通则》（CXS 283-1978）第7.2条所规定的特征描述用语或是符合《营养和保健声明使用准则》（CXG 23-1997）[1]的营养声明。

该名称也可用于符合本标准规定的干酪做成的切块、切片、切碎或磨碎产品。

7.2 原产国

应当标记原产国（即生产国，而非名称起源国）。当产品在第二个国家有实质性改造[2]，则以标识目的进行改造的国家应视为原产国。

7.3 乳脂含量说明

乳脂含量应以销售国可接受的方式声明：（i）质量百分比形式；（ii）干物质中脂肪百分比形式；（iii）如果产品标识上标明了份数，也可用每份中乳脂重量（g）表示。

7.4 非零售包装标识

在包装容器上除要标明产品名称、批次、生产厂家或包装商的名称和地址外，在包装容器上或附随说明书中，也应按本标准第7条和《预包装食品标识通用标准》（CXS 1-1985）第4.1~4.8条所要求的信息以及储存说明（在必要的地方）加以陈述。如果批次、生产厂家或包装商的名称和地址可以在附随的文件标明，则可以用一个识别标识来代替。

8 抽样和分析方法

为检查产品是否符合本标准，应采用《分析和抽样推荐性方法》（CXS 234-1999）中涉及本标准规定的分析和抽样方法。

---

1 为比较营养声明，干物质中脂肪含量最低参考值为40%。
2 例如，重新包装、切块、切片、切碎和磨碎不构成实质性改变。

附录

附加信息

以下补充信息对前文各条款的规定不构成影响，前文规定内容对产品标识、食品名称的使用以及食品安全性至关重要。

1 加工方法

1.1 发酵过程：微生物产酸发酵。

1.2 凝固类型：乳蛋白凝固通常是合适凝固温度下微生物酸化和蛋白酶（如凝乳酶）的共同作用结果。

STANDARD FOR COULOMMIERS

CXS 274-1969

Formerly CODEX STAN C-18-1969. Adopted in 1969. Revised in 2007.

Amended in 2008, 2010, 2018, 2019.

1 SCOPE

This Standard applies to Coulommiers intended for direct consumption or for further processing in conformity with the description in Section 2 of this Standard.

2 DESCRIPTION

Coulommiers is a soft, surface ripened, primarily mould ripened cheese in conformity with the *General Standard for Cheese* (CXS 283-1978) which has a shape of a flat cylinder or sectors thereof. The body has a near white through to light yellow colour and a soft-textured (when pressed by thumb), but not crumbly texture, ripened from the surface to the center of the cheese. Gas holes are generally absent, but few openings and splits are acceptable. A rind is to be developed that is soft and entirely covered with white mould but may have red, brownish or orange coloured spots. Whole cheese may be cut or formed into sectors prior to or after the mould development.

For Coulommiers ready for consumption, the ripening procedure to develop flavour and body characteristics is normally from 10 days at 10–16 °C depending on the extent of maturity required. Alternative ripening conditions (including the addition of ripening enhancing enzymes) may be used, provided the cheese exhibits similar physical, biochemical and sensory properties as those achieved by the previously stated ripening procedure. Coulommiers intended for further processing need not exhibit the same extent of ripening when justified through technical and/or trade needs.

3 ESSENTIAL COMPOSITION AND QUALITY FACTORS

3.1 Raw materials

Cows' milk or buffaloes' milk, or their mixtures, and products obtained from these milks.

3.2 Permitted ingredients

- Starter cultures of harmless lactic acid and/or flavour producing bacteria and cultures of other harmless micro-organisms, including *Geotrichum candidum*, *Brevibacterium linens*, and yeast;
- Rennet or other safe and suitable coagulating enzymes;
- Sodium chloride and potassium chloride as a salt substitute;
- Potable water;
- Safe and suitable processing aids;
- Safe and suitable enzymes to enhance the ripening process;
- Rice, corn and potato flours and starches: Notwithstanding the provisions in the *General Standard for Cheese* (CXS 283-1978), these substances can be used in

the same function as anti-caking agents for treatment of the surface of cut, sliced, and shredded products only, provided they are added only in amounts functionally necessary as governed by Good Manufacturing Practice, taking into account any use of the anti-caking agents listed in Section 4.

3.3 Composition

Milk constituent	Minimum content (m/m)	Maximum content (m/m)	Reference level
Milkfat in dry matter	40%	Not restricted	40% to 50%
Dry matter	Depending on the fat in dry matter content, according to the table below			

Fat in dry matter content (m/m)	Corresponding minimum dry matter content (m/m)
Equal to or above 40% but less than 50%	42%
Equal to or above 50% but less than 60%	46%
Equal to or above 60%	52%

Compositional modifications beyond the minima and maxima specified above for milkfat and dry matter are not considered to be in compliance with Section 4.3.3 of the *General Standard for the Use of Dairy Terms* (CXS 206-1999).

3.4 Essential sizes and shapes

Maximum height: Approx. 5 cm;

Weight: Whole cheese of flat cylinder: min. 300 g.

3.5 Essential ripening procedure

Rind formation and maturation (proteolysis) from the surface to the center is predominantly caused by *Penicillium candidum* and/or *Penicillium camembertii* and *Penicillium caseicolum*.

4 FOOD ADDITIVES

Only those additives classes indicated as justified in the table below may be used for the product categories specified. Colours used in accordance with Tables 1 and 2 of the *General Standard for Food Additives* (CXS 192-1995) in food category 01.6.2.1 (Ripened cheese, includes rind) and only certain acidity regulators in Table 3 are acceptable for use in foods conforming to this standard.

Additive functional class	Justified use	
	Cheese mass	Surface/rind treatment
Colours	×[a]	–
Bleaching agents	–	–
Acids	–	–
Acidity regulators	×	–

Additive functional class	Justified use	
	Cheese mass	Surface/rind treatment
Stabilizers	–	–
Thickeners	–	–
Emulsifiers	–	–
Antioxidants	–	–
Preservatives	–	–
Foaming agents	–	–
Anti–caking agents	–	–

(a) Only to obtain the colour characteristics, as described in Section 2.

× The use of additives belonging to the class is technologically justified.

– The use of additives belonging to the class is not technologically justified.

5 CONTAMINANTS

The products covered by this Standard shall comply with the maximum levels for contaminants that are specified for the product in the *General Standard for Contaminants and Toxins in Food and Feed* (CXS 193-1995).

The milk used in the manufacture of the products covered by this Standard shall comply with the maximum levels for contaminants and toxins specified for milk by the *General Standard for Contaminants and Toxins in Food and Feed* (CXS 193-1995) and with the maximum residue limits for veterinary drug residues and pesticides established for milk by the CAC.

6 HYGIENE

It is recommended that the product covered by the provisions of this Standard be prepared and handled in accordance with the appropriate sections of the *General Principles of Food Hygiene* (CXC 1-1969), the *Code of Hygienic Practice for Milk and Milk Products* (CXC 57-2004) and other relevant Codex texts such as Codes of Hygienic Practice and Codes of Practice. The products should comply with any microbiological criteria established in accordance with the *Principles and Guidelines for the Establishment and Application of Microbiological Criteria Related to Foods* (CXG 21-1997).

7 LABELLING

In addition to the provisions of the *General Standard for the Labelling of Prepackaged Foods* (CXS 1-1985) and the *General Standard for the Use of Dairy Terms* (CXS 206-1999), the following specific provisions apply.

7.1 Name of the food

The name Coulommiers may be applied in accordance with Section 4.1 of the *General Standard for the Labelling of Prepackaged Foods* (CXS 1-1985), provided that the product is in conformity with this Standard. Where customary in the country of retail sale, alternative spelling may be used.

The use of the name is an option that may be chosen only if the cheese complies with this Standard. Where the name is not used for a cheese that complies with this Standard, the naming provisions of the *General Standard for Cheese* (CXS 283-1978) apply.

The designation of products in which the fat content is above the reference range specified in Section 3.3 of this Standard shall be accompanied by an appropriate qualification describing the modification made or the fat content (expressed as fat in dry matter or as percentage by mass whichever is acceptable in the country of retail sale), either as part of the name or in a prominent position in the same field of vision. Suitable qualifiers are the appropriate characterizing terms specified in Section 7.2 of the *General Standard for Cheese* (CXS 283-1978) or a nutritional claim in accordance with the *Guidelines for Use of Nutrition and Health Claims* (CXG 23-1997)[1].

The designation may also be used for cut, sliced, shredded or grated products made from cheese which cheese is in conformity with this Standard.

7.2 Country of origin

The country of origin (which means the country of manufacture, not the country in which the name originated) shall be declared. When the product undergoes substantial transformation[2] in a second country, the country in which the transformation is performed shall be considered to be the country of origin for the purpose of labelling.

7.3 Declaration of milkfat content

The milk fat content shall be declared in a manner found acceptable in the country of retail sale either (i) as a percentage by mass, (ii) as a percentage of fat in dry matter, or (iii) in grams per serving as quantified in the label, provided that the number of servings is stated.

7.4 Labelling of non-retail containers

Information specified in Section 7 of this Standard and Sections 4.1 to 4.8 of the *General Standard for the Labelling of Prepackaged Foods* (CXS 1-1985) and, if necessary, storage instructions, shall be given either on the container or in accompanying documents, except that the name of the product, lot identification, and the name of the manufacturer or packer shall appear on the container, and in the absence of such a container, on the product itself. However, lot identification and the name and address may be replaced by an identification mark, provided that such mark is clearly identifiable with the accompanying documents.

8 METHODS OF SAMPLING AND ANALYSIS

For checking the compliance with this Standard, the methods of analysis and sampling contained in the *Recommended Methods of Analysis and Sampling* (CXS 234-1999) relevant to the provisions in this Standard, shall be used.

---

1   For the purpose of comparative nutritional claims, the minimum fat content of 40% fat in dry matter constitutes the reference.

2   For instance, repackaging, cutting, slicing, shredding and grating is not regarded as substantial transformation.

APPENDIX

ADDITIONAL INFORMATION

The additional information below does not affect the provisions in the preceding sections which are those that are essential to the product identity, the use of the name of the food and the safety of the food.

1 **Method of manufacture**

1.1 Fermentation procedure: Microbiologically derived acid development.

1.2 Type of coagulation: Coagulation of the milk protein is typically obtained through the combined action of microbial acidification and proteases (e.g. rennet) at an appropriate coagulation temperature.

奶 油 干 酪

STANDARD FOR CREAM CHEESE

CXS 275-1973

前为CODEX STAN C-31-1973。1973年通过。2007年修订。

2008年、2010年、2016年、2018年修正。

1 范围

本标准适用于符合本标准第2条所述即食或用于进一步加工的奶油干酪。

在一些国家，"奶油干酪"一词用以指代不符合第2条说明的干酪，例如高脂成熟硬质干酪。本标准不适用于此类干酪。

2 说明

奶油干酪是指符合《未成熟干酪（包括新鲜干酪）标准》（CXS 221-2001）和《干酪通则》（CXS 283-1978）的软质、涂抹状、未成熟和无外皮[1]干酪。这种干酪颜色介于近白色与淡黄色之间。质地呈涂抹状，介于光滑与略成片状之间，无孔洞，可涂抹于其他食品，也可与其他食品直接混食。

3 基本成分和质量要求

3.1 基本原料

乳和/或乳制品。

3.2 其他配料

- 无害乳酸发酵剂和/或产香菌和其他无害微生物的培养剂；
- 凝乳酶或其他安全、适宜的凝固酶；
- 作为代盐制品的氯化钠和氯化钾；
- 饮用水；
- 安全、适宜加工助剂；
- 明胶和淀粉：用来发挥相当于稳定剂功能，但添加剂量仅以发挥必要功能为限，要以考虑使用第4条所列稳定剂或增稠剂的良好生产规范为准；
- 醋。

---

[1] 这种干酪在保存中不会结成外皮（"无外皮"干酪）。

3.3 成分

乳成分	最低含量（m/m）	最高含量（m/m）	参考含量（m/m）
干物质乳脂	25%	/	60%~70%
脱脂水分	67%	—	未定
干物质	22%	以脱脂水分为限	未定

乳脂和干物质含量超过规定的最大值和最小值被视为不符合《乳品术语》（CXS 206-1999）第4.3.3条。

4 食品添加剂

只有下表所示可合理使用的添加剂类别才可用于指定产品类别。在允许使用的各类添加剂中，只可使用下表所列的食品添加剂，并以规定的功能和限量为限。

添加剂功能类别	合理使用	
	干酪块	表面/外皮处理
着色剂	×(a)	—
漂白剂	—	—
酸度调节剂	×	—
稳定剂	×(b)	—
增稠剂	×(b)	—
乳化剂	×	—
抗氧化剂	×	—
防腐剂	×(b)	—
发泡剂	×(c)	—
抗结块剂	—	—

（a）仅为获得第2条所述颜色特征。
（b）可按乳制品定义使用包括改性淀粉在内的稳定剂和增稠剂，但仅以发挥必要功能为限，用于加热经处理的产品，并考虑使用第3.2条所述明胶和淀粉。
（c）仅用于打发产品。
× 使用该类别添加剂符合技术要求。
— 使用该类别添加剂不符合技术要求。

INS编号	添加剂名称	最大限量
防腐剂		
200	山梨酸	1 000 mg/kg 单用或混用，以山梨酸计
202	山梨酸钾	
203	山梨酸钙	

INS编号	添加剂名称	最大限量
234	乳酸链球菌素	12.5 mg/kg
280	丙酸	根据GMP限量使用
281	丙酸钠	
282	丙酸钙	
283	丙酸钾	
酸度调节剂		
170（i）	碳酸钙	根据GMP限量使用
260	冰醋酸	
261（i）	醋酸钾	
261（ii）	双乙酸钾	
262（i）	乙酸钠	
263	乙酸钙	
270	乳酸（*L-*，*D-*和*DL-*）	
296	*DL*-苹果酸	
325	乳酸钠	
326	乳酸钾	
327	乳酸钙	
330	柠檬酸	
331（i）	柠檬酸二氢钠	
332（i）	柠檬酸二氢钾	
333	柠檬酸钙	
334	*L*(+)-酒石酸	1 500 mg/kg 单用或混用，以酒石酸计
335（ii）	*L*(+)-酒石酸钠	
337	*L*(+)-酒石酸钾钠	
338	磷酸	880 mg/kg，以磷计
350（i）	*DL*-苹果酸氢钠	根据GMP限量使用
350（ii）	*DL*-苹果酸钠	
352（ii）	*D*，*L*-苹果酸钙	
500（i）	碳酸钠	

INS编号	添加剂名称	最大限量
500（ii）	碳酸氢钠	根据GMP限量使用
500（iii）	倍半碳酸钠	
501（i）	碳酸钾	
501（ii）	碳酸氢钾	
504（i）	碳酸镁	
504（ii）	碳酸氢镁	
507	盐酸	
575	葡萄糖酸-δ-内酯	
577	葡萄糖酸钾	
578	葡萄糖酸钙	
稳定剂		
339（i）	磷酸二氢钠	4 400 mg/kg 单用或混用，以磷计
339（ii）	磷酸氢二钠	
339（iii）	磷酸三钠	
340（i）	磷酸二氢钾	
340（ii）	磷酸氢二钾	
340（iii）	磷酸三钾	
341（i）	磷酸二氢钙	
341（ii）	磷酸氢二钙	
341（iii）	磷酸三钙	
342（i）	磷酸二氢铵	
342（ii）	磷酸氢二铵	
343（ii）	磷酸氢二镁	
343（iii）	磷酸三镁	
450（i）	磷酸二钠	
450（iii）	焦磷酸四钠	
450（v）	二磷酸四钾	
450（vi）	磷酸二钙	
451（i）	三磷酸五钠	

INS编号	添加剂名称	最大限量
451（ii）	三磷酸五钾	4 400 mg/kg 单用或混用，以磷计
452（i）	六偏磷酸钠	
452（ii）	聚磷酸钾	
452（iv）	聚磷酸钙	
452（v）	聚磷酸铵	
400	海藻酸	根据GMP限量使用
401	海藻酸钠	
402	海藻酸钾	
403	海藻酸铵	
404	海藻酸钙	
405	海藻酸丙二醇酯	5 000 mg/kg
406	琼脂	根据GMP限量使用
407	卡拉胶	
407a	加工琼芝属海藻胶（PES）	
410	槐豆胶（又名刺槐豆胶）	
412	瓜尔胶	
413	黄蓍胶	
415	黄原胶	
416	刺梧桐胶	
417	刺云实胶	
418	结冷胶	
466	羧甲基纤维素钠（纤维素胶）	
1400	糊精，焙炒淀粉	
1401	酸处理淀粉	
1402	碱处理淀粉	
1403	漂白淀粉	
1404	氧化淀粉	
1405	酶处理淀粉	
1410	单淀粉磷酸酯	

INS编号	添加剂名称	最大限量
1412	磷酸酯双淀粉	根据GMP限量使用
1413	磷酸化二淀粉磷酸酯	
1414	乙酰化二淀粉磷酸酯	
1420	醋酸酯淀粉	
1422	乙酰化双淀粉己二酸酯	
1440	羟丙基淀粉	
1442	羟丙基二淀粉磷酸酯	
乳化剂		
322	卵磷脂	根据GMP限量使用
470（i）	肉豆蔻酸、棕榈酸和硬脂酸与氨、钙、钾和钠的盐	
470（ii）	油酸与钙、钾、钠的盐	
471	单、双甘油脂肪酸酯	
472a	乙酰化单、双甘油脂肪酸酯	
472b	乳酸脂肪酸甘油酯	
472c	柠檬酸脂肪酸甘油酯	
472e	二乙酰酒石酸酯和脂肪酸酯	10 000 mg/kg
抗氧化剂		
300	L-抗坏血酸	根据GMP限量使用
301	抗坏血酸钠	
302	抗坏血酸钙	
304	抗坏血酸棕榈酸酯	500 mg/kg 单用或混用，以抗坏血酸硬脂酸酯计
305	抗坏血酸硬脂酸酯	
307b	生育酚浓缩物（混合物）	200 mg/kg 单用或混用
307c	DL-α-生育酚	
着色剂		
160a（i）	β-胡萝卜素（人工合成）	35 mg/kg，单独或混合使用
160a（iii）	β-胡萝卜素（三孢布拉霉）	
160e	β-apo-8'-胡萝卜素	
160f	β-apo-8'-胡萝卜素酸乙酯	

INS编号	添加剂名称	最大限量
160a（ii）	β-胡萝卜素（天然萃取物）	600 mg/kg
160b（ii）	胭脂树提取物-红木素	25 mg/kg
171	二氧化钛	根据GMP限量使用
发泡剂		
290	二氧化碳	根据GMP限量使用
941	氮气	

5　污染物

本标准所涉及的产品应符合《食品及饲料中污染物和毒素通用标准》（CXS 193-1995）规定的污染物最大限量。

本标准所涉及的产品在加工中使用的牛乳应符合《食品及饲料中污染物和毒素通用标准》（CXS 193-1995）规定的乳中污染物和毒素最大限量以及国际食品法典委员会设定的农药和兽药最大残留限量。

6　卫生要求

建议本标准所涉及的产品应遵循《食品卫生总则》（CXC 1-1969）、《乳及乳制品卫生操作规范》（CXC 57-2004）以及《卫生操作规范》和《生产操作规范》等其他相关法典文本。产品应符合《食品微生物标准制定与实施原则和准则》（CXG 21-1997）规定的所有微生物标准。

7　标识

除符合《预包装食品标识通用标准》（CXS 1-1985）和《乳品术语》（CXS 206-1999）外，还应符合以下具体规定。

7.1　产品名称

奶油干酪名称应符合《预包装食品标识通用标准》（CXS 1-1985）第4.1条，只要产品符合该标准，产品销售国的习惯拼写也可被使用。

只要干酪符合本标准，就可选择使用此名称。对符合本标准但不使用此名称的干酪，应按《干酪通则》（CXS 283-1978）命名规定来命名。

对于脂肪含量低于或高于参考范围但相等于或高于本标准第3.3条规定的干物质脂肪含量40%的产品，标示时应附上相应限定语，说明所作改动或脂肪含量（以干物质所含脂肪或销售国接受的各种质量百分比表示），作为名称的一部分或处于与名称同一视野内的显眼位置。对于脂肪含量低于40%的干物质脂肪但高于本标准第3.3条规定的绝对最低含量产品，可以两种方式予以标示：附上相应修饰语，说明所作改动或脂肪含量（以干物质所含脂肪或质量百分比表示），作为名称的一部分或处于与名称同一视野内的显眼位置；或采用产品生产国和/或销售国国内立法指定的名称，或加注现行通用名称。但无论是哪种情况，所用标示方式均不得在零售中使消费者对干酪特点和特性产生误解。

适当修饰语是指《干酪通则》（CXS 283-1978）第7.2条规定的特征描述用语或是符合《营养和保健声明使用准则》（CXG 23-1997）的营养声明[1]。

7.2 原产国

应当标记原产国（即生产国，而非名称起源国）。当产品在第二个国家有实质性改造[2]，则以标识目的进行改造的国家应视为原产国。

7.3 乳脂含量说明

乳脂含量应以销售国可接受的方式声明：（i）质量百分比形式；（ii）干物质中脂肪百分比形式；（iii）如果产品标识上标明了份数，也可用每份中乳脂重量（g）表示。

7.4 非零售包装标识

在包装容器上除要标明产品名称、批次、生产厂家或包装商的名称和地址外，在包装容器上或附随说明书中，也应按本标准第7条和《预包装食品标识通用标准》（CXS 1-1985）第4.1～4.8条所要求的信息以及储存说明（在必要的地方）加以陈述。如果批次、生产厂家或包装商的名称和地址可以在附随的文件标明，则可以用一个识别标识来代替。

8 抽样和分析方法

为检查产品是否符合本标准，应采用《分析和抽样推荐性方法》（CXS 234-1999）中涉及本标准规定的分析和抽样方法。

---

1 为比较营养声明，干物质中脂肪含量最低参考值为60%。
2 例如，重新包装、切块、切片、切碎和磨碎不构成实质性改变。

STANDARD FOR CREAM CHEESE

CXS 275-1973

Formerly CODEX STAN C-31-1973. Adopted in 1973. Revised in 2007.

Amended in 2008, 2010, 2016, 2018.

1 **SCOPE**

This Standard applies to Cream Cheese intended for direct consumption or for further processing in conformity with the description in Section 2 of this Standard.

In some countries, the term "cream cheese" is used to designate cheeses, such as high fat ripened hard cheese, that do not conform to the description in Section 2. This Standard does not apply to such cheeses.

2 **DESCRIPTION**

Cream Cheese is a soft, spreadable, unripened and rindless[1] cheese in conformity with the *Group Standard for Unripened Cheese Including Fresh Cheese* (CXS 221-2001) and the *General Standard for Cheese* (CXS 283-1978). The cheese has a near white through to light yellow colour. The texture is spreadable and smooth to slightly flaky and without holes, and the cheese spreads and mixes readily with other foods.

3 **ESSENTIAL COMPOSITION AND QUALITY FACTORS**

3.1 **Raw materials**

Milk and/or products obtained from milk.

3.2 **Permitted ingredients**

- Starter cultures of harmless lactic acid and/or flavour producing bacteria and cultures of other harmless micro-organisms;

- Rennet or other safe and suitable coagulating enzymes;

- Sodium chloride and potassium chloride as a salt substitute;

- Potable water;

- Safe and suitable processing aids;

- Gelatine and starches: These substances can be used in the same function as stabilizers, provided they are added only in amounts functionally necessary as governed by Good Manufacturing Practice taking into account any use of the stabilizers/thickeners listed in Section 4;

- Vinegar.

---

1 The cheese has been kept in such a way that no rind is developed (a "rindless" cheese).

3.3 Composition

Milk constituent	Minimum content (m/m)	Maximum content (m/m)	Reference level (m/m)
Milk fat in dry matter	25%	Not restricted	60%–70%
Moisture on fat free basis	67%	–	Not specified
Dry matter	22%	Restricted by the MFFB	Not specified

Compositional modifications of Cream Cheese beyond the minima and maxima specified above for milkfat, moisture and dry matter are not considered to be in compliance with Section 4.3.3 of the *General Standard for the Use of Dairy Terms* (CXS 206-1999).

4 FOOD ADDITIVES

Only those additives classes indicated as justified in the table below may be used for the product categories specified. Within each additive class, and where permitted according to the table, only those food additives listed below may be used and only within the functions and limits specified.

Additive functional class	Justified use	
	Cheese mass	Surface/rind treatment
Colours	×[a]	–
Bleaching agents	–	–
Acidity regulators	×	–
Stabilizers	×[b]	–
Thickeners	×[b]	–
Emulsifiers	×	–
Antioxidants	×	–
Preservatives	×[b]	–
Foaming agents	×[c]	–
Anticaking agents	–	–

(a) Only to obtain the colour characteristics, as described in Section 2.

(b) Stabilizers and thickeners including modified starches may be used in compliance with the definition of milk products and only to heat treated products to the extent they are functionally necessary, taking into account any use of gelatine and starches as provided for in Section 3.2.

(c) For whipped products, only.

× The use of additives belonging to the class is technologically justified.

– The use of additives belonging to the class is not technologically justified.

INS no.	Name of additive	Maximum level
Preservatives		
200	Sorbic acid	1 000 mg/kg, singly or in combination as sorbic acid
202	Potassium sorbate	
203	Calcium sorbate	
234	Nisin	12.5 mg/kg
280	Propionic acid	Limited by GMP
281	Sodium propionate	
282	Calcium propionate	
283	Potassium propionate	
Acidity regulators		
170(i)	Calcium carbonate	Limited by GMP
260	Acetic acid, glacial	
261(i)	Potassium acetate	
261(ii)	Potassium diacetate	
262(i)	Sodium acetate	
263	Calcium acetate	
270	Lactic acid, *L*-, *D*- and *DL*-	
296	Malic acid, *DL*-	
325	Sodium lactate	
326	Potassium lactate	
327	Calcium lactate	
330	Citric acid	
331(i)	Sodium dihydrogen citrate	
332(i)	Potassium dihydrogen citrate	
333	Calcium citrates	
334	Tartaric acid, *L*(+)-	1 500 mg/kg, singly or in combination as tartaric acid
335(ii)	Sodium *L*(+)-tartrate	
337	Potassium sodium *L*(+)-tartrate	
338	Phosphoric acid	880 mg/kg as phosporous
350(i)	Sodium hydrogen *DL*-malate	Limited by GMP
350(ii)	Sodium *DL*-malate	
352(ii)	Calcium malate, *D*, *L*-	
500(i)	Sodium carbonate	

INS no.	Name of additive	Maximum level
500(ii)	Sodium hydrogen carbonate	Limited by GMP
500(iii)	Sodium sesquicarbonate	
501(i)	Potassium carbonate	
501(ii)	Potassium hydrogen carbonate	
504(i)	Magnesium carbonate	
504(ii)	Magnesium hydrogen carbonate	
507	Hidrochloric acid	
575	Glucono-*delta*-lactone	
577	Potassium gluconate	
578	Calcium gluconate	
Stabilizers		
339(i)	Sodium dihydrogen phosphate	4 400 mg/kg singly or in combination, expressed as phosphorus
339(ii)	Disodium hydrogen phosphate	
339(iii)	Trisodium phosphate	
340(i)	Potassium dihydrogen phosphate	
340(ii)	Dipotassium hydrogen phosphate	
340(iii)	Tripotassium phosphate	
341(i)	Calcium dihydrogen phosphate	
341(ii)	Calcium hydrogen phosphate	
341(iii)	Tricalcium orthophosphate	
342(i)	Ammonium dihydrogen phosphate	
342(ii)	Diammonium hydrogen phosphate	
343(ii)	Magnesium hydrogen phosphate	
343(iii)	Trimagnesium phosphate	
450(i)	Disodium diphosphate	
450(iii)	Tetrasodium diphosphate	
450(v)	Tetrapotassium diphosphate	
450(vi)	Dicalcium diphosphate	
451(i)	Pentasodium triphosphate	
451(ii)	Pentapotassium triphosphate	
452(i)	Sodium polyphosphate	
452(ii)	Potassium polyphosphate	
452(iv)	Calcium polyphosphate	

INS no.	Name of additive	Maximum level
452(v)	Ammonium polyphosphate	4 400 mg/kg singly or in combination, expressed as phosphorus
400	Alginic acid	Limited by GMP
401	Sodium alginate	
402	Potassium alginate	
403	Ammonium alginate	
404	Calcium alginate	
405	Propylene glycol alginate	5 000 mg/kg
406	Agar	Limited by GMP
407	Carrageenan	
407a	Processed euchema seaweed (PES)	
410	Carob bean gum	
412	Guar gum	
413	Tragacanth gum	
415	Xanthan gum	
416	Karaya gum	
417	Tara gum	
418	Gellan gum	
466	Sodium carboxymethyl cellulose (Cellulose gum)	
1400	Dextrins, roasted starch	
1401	Acid-treated starch	
1402	Alkaline treated starch	
1403	Bleached starch	
1404	Oxidized starch	
1405	Starches, enzyme-treated	
1410	Monostarch phosphate	
1412	Distarch phosphate	
1413	Phosphateddistarch phosphate	
1414	Acetylated distarch phosphate	
1420	Starch acetate	
1422	Acetylated distarchadipate	
1440	Hydroxypropyl starch	

INS no.	Name of additive	Maximum level
1442	Hydroxypropyldistarch phosphate	Limited by GMP
Emulsifiers		
322	Lecithins	Limited by GMP
470(i)	Salt of myristic, palmitic and stearic acids wit ammonia, calcium, potassium and sodium	
470(ii)	Salt of oleic acid with calcium, potassium and sodium	
471	Mono- and di-glycerides of fatty acids	
472a	Acetic and fatty acid esters of glycerol	
472b	Lactic and fatty acid esters of glycerol	
472c	Citric and fatty acid esters of glycerol	
472e	Diacetyltartaric and fatty acid esters of glycerol	10 000 mg/kg
Antioxidants		
300	Ascorbic acid, *L*-	Limited by GMP
301	Sodium ascorbate	
302	Calcium ascorbate	
304	Ascorbyl palmitate	500 mg/kg singly or in combination as ascorbyl stearate
305	Ascorbyl stearate	
307b	Tocopherol concentrate, mixed	200 mg/kg singly or in combination
307c	Tocopherol, dl-*alpha*-	
Colours		
160a(i)	Carotene, *beta*-, synthetic	35 mg/kg, singly or in combination
160a(iii)	Carotene , *beta*-, *Blakesleatrispora*	
160e	Carotenal, *beta-apo*-8'-	
160f	Carotenoicacid, ethylester, *beta*-apo-8'-	
160a(ii)	Carotenes, *beta*-, vegetable	600 mg/kg
160b(ii)	Annatto extracts-norbixin based	25 mg/kg
171	Titanium dioxide	Limited by GMP
Foaming agent		
290	Carbon dioxide	Limited by GMP
941	Nitrogen	

5 CONTAMINANTS

The products covered by this Standard shall comply with the maximum levels for contaminants that are specified for the product in the *General Standard for Contaminants and Toxins in Food and Feed* (CXS 193-1995).

The milk used in the manufacture of the products covered by this Standard shall comply with the maximum levels for contaminants and toxins specified for milk by the *General Standard for Contaminants and Toxins in Food and Feed* (CXS 193-1995) and with the maximum residue limits for veterinary drug residues and pesticides established for milk by the CAC.

6 HYGIENE

It is recommended that the product covered by the provisions of this Standard be prepared and handled in accordance with the appropriate sections of the *General Principles of Food Hygiene* (CXC 1-1969), the *Code of Hygienic Practice for Milk and Milk Products* (CXC 57-2004) and other relevant Codex texts such as Codes of Hygienic Practice and Codes of Practice.

The products should comply with any microbiological criteria established in accordance with the *Principles and Guidelines for the Establishment and Application of Microbiological Criteria Related to Foods* (CXG 21-1997).

7 LABELLING

In addition to the provisions of the *General Standard for the Labelling of Prepackaged Foods* (CXS 1-1985) and the *General Standard for the Use of Dairy Terms* (CXS 206-1999), the following specific provisions apply.

7.1 Name of the food

The name Cream Cheese may be applied in accordance with Section 4.1 of the *General Standard for the Labelling of Prepackaged Foods* (CXS 1-1985), provided that the product is in conformity with this Standard. Where customary in the country of retail sale, alternative spelling may be used. The name may be translated into other languages so that the consumer in the country of retail sale will not be misled.

The use of the name is an option that may be chosen only if the cheese complies with this Standard. Where the name is not used for a cheese that complies with this Standard, the naming provisions of the *General Standard for Cheese* (CXS 283-1978) apply.

The designation of products in which the fat content is below or above the reference range but equal to or above 40% fat in dry matter as specified in Section 3.3 of this Standard shall be accompanied by an appropriate qualification describing the modification made or the fat content (expressed as fat in dry matter or as percentage by mass whichever is acceptable in the country of retail sale), either as part of the name or in a prominent position in the same field of vision. The designation of products in which the fat content is below 40% fat in dry matter but above the absolute minimum specified in Section 3.3 of this Standard shall *either* be accompanied by an appropriate qualifier describing the modification made or the fat content (expressed as fat in dry matter or as percentage by mass), either as part of the name or in a prominent position in the same field of vision, *or alternatively* the name specified in the national legislation of the country in which the product is manufactured and/or sold or with a name existing by common usage, in either case provided that the designation used does not create an erroneous impression the retail sale regarding the character and identity of the cheese.

Suitable qualifiers are the appropriate characterizing terms specified in Section 7.2 of the *General Standard for Cheese* (CXS 283-1978) or a nutritional claim in accordance with the *Guidelines for Use of Nutrition and Health Claims* (CXG 23-1997)[1].

7.2 Country of origin

The country of origin (which means the country of manufacture, not the country in which the name originated) shall be declared. When the product undergoes substantial transformation[2] in a second country, the country in which the transformation is performed shall be considered to be the country of origin for the purpose of labelling.

7.3 Declaration of milkfat content

The milk fat content shall be declared in a manner found acceptable in the country of retail sale, either (i) as a percentage by mass, (ii) as a percentage of fat in dry matter, or (iii) in grams per serving as quantified in the label, provided that the number of servings is stated.

7.4 Labelling of non-retail containers

Information specified in Section 7 of this Standard and Sections 4.1 to 4.8 of the *General Standard for the Labelling of Prepackaged Foods* (CXS 1-1985) and, if necessary, storage instructions, shall be given either on the container or in accompanying documents, except that the name of the product, lot identification, and the name of the manufacturer or packer shall appear on the container, and in the absence of such a container, on the product itself. However, lot identification and the name and address may be replaced by an identification mark, provided that such mark is clearly identifiable with the accompanying documents.

8 METHODS OF SAMPLING AND ANALYSIS

For checking the compliance with this Standard, the methods of analysis and sampling contained in the *Recommended Methods of Analysis and Sampling* (CXS 234-1999) relevant to the provisions in this Standard, shall be used.

---

[1] For the purpose of comparative nutritional claims, the minimum fat content of 60% fat in dry matter constitutes the reference.

[2] For instance, repackaging, cutting, slicing, shredding and grating is not regarded as substantial transformation.

卡门培尔干酪

STANDARD FOR CAMEMBERT

CXS 276-1973

前为CODEX STAN C-33-1973。1973年通过。2007年修订。
2008年、2010年、2018年、2019年修正。

1　范围

本标准适用于符合本标准第2条所述即食或用于进一步加工的卡门培尔干酪。

2　说明

卡门培尔干酪是一种软质表面成熟干酪，主要属于《干酪通则》（CXS 283-1978）中提及的霉菌成熟干酪，形状呈扁圆柱形或扇形。通体具有淡黄透白色，质地柔软（当拇指摁压时）但不易碎，从表面到中心完全成熟。几乎无气孔，但允许个别开口或裂口。干酪带有软质且完全被白霉覆盖的外皮，可有红色、褐色或橙色斑点。完整干酪在霉菌生成前后可切分或制成扇形。

对于即食的卡门培尔干酪，调制味道和外观特征的催熟程序在10～16 ℃下通常需要10天。在干酪具备和前述的催熟程序同样的物理、生化和感官特性的情况下，可以使用其他催熟条件（包括添加催熟的酶）。如有合理技术或贸易需要，用于进一步加工的卡门培尔干酪，无需具备同等成熟度。

卡雷德卡门培尔干酪是一种方形柔软表面成熟的干酪，符合卡门培尔干酪的其他标准和要求。

3　基本成分和质量要求

3.1　基本原料

牛乳、水牛乳或它们的混合乳、以及源于这些乳的制品。

3.2　其他配料

- 无害乳酸发酵剂和/或产香菌和其他无害微生物的培养剂，包括白地霉，亚麻短杆菌和酵母菌；
- 凝乳酶或其他安全、适宜的凝固酶；
- 作为代盐制品的氯化钠和氯化钾；
- 饮用水；
- 促进成熟的安全、适宜的酶；

- 安全、适宜的加工剂；
- 大米、玉米、马铃薯粉和淀粉：尽管有《干酪通则》（CXS 283-1978），但这些物质同样可作为抗结块剂处理切块、切片和切碎产品表面，只要添加量是良好操作规范所要求达到期待功能的必需量，同时考虑第4条规定的抗结块剂的使用。

3.3 成分

乳成分	最低含量（m/m）	最高含量（m/m）	参考含量（m/m）
干物质乳脂	30%	/	45%~55%
干物质	根据干物质脂肪含量，参见下表		

干物质脂肪含量（m/m）	相应最低干物质含量（m/m）
≥30%，<40%	38%
≥40%，<45%	41%
≥45%，<55%	43%
≥55%	48%

乳脂和干物质含量超过规定的最大值和最小值被视为不符合《乳品术语》（CXS 206-1999）第4.3.3条。

3.4 基本尺寸和形状

最大高度：大约5 cm；

重量：整个平柱形或方形干酪：80~500 g。

3.5 基本成熟过程

外皮的形成和从表面到中心的成熟（蛋白质水解）主要在白青霉和/或卡门培尔青霉和白酪青霉作用下促成。

4 食品添加剂

只有下表所示可以合理使用的添加剂类别才可用于指定产品类别。根据《食品添加剂通用标准》（CXS 192-1995）表1和表2在食品类别01.6.2.1（成熟干酪，包括表皮）中使用的着色剂，以及表3中仅部分酸度调节剂可用于符合本标准的食品。

添加剂功能类别	合理使用	
	干酪块	表面/外皮处理
着色剂	×[a]	−
漂白剂	−	−

添加剂功能类别	合理使用	
	干酪块	表面/外皮处理
酸	-	-
酸度调节剂	×	-
稳定剂	-	-
增稠剂	-	-
乳化剂	-	-
抗氧化剂	-	-
防腐剂	-	-
发泡剂	-	-
抗结块剂	-	-

（a）仅为获得第2条所述颜色特征。

× 使用该类别添加剂符合技术要求。

− 使用该类别添加剂不符合技术要求。

5 污染物

本标准所涉及的产品应符合《食品及饲料中污染物和毒素通用标准》（CXS 193-1995）规定的污染物最大限量。

本标准所涉及的产品在加工中使用的牛乳应符合《食品及饲料中污染物和毒素通用标准》（CXS 193-1995）规定的乳中污染物和毒素最大限量以及国际食品法典委员会设定的农药和兽药最大残留限量。

6 卫生要求

建议本标准所涉及的产品应遵循《食品卫生总则》（CXC 1-1969）、《乳及乳制品卫生操作规范》（CXC 57-2004）以及《卫生操作规范》和《生产操作规范》等其他相关法典文本。产品应符合《食品微生物标准制定与实施原则和准则》（CXG 21-1997）规定的所有微生物标准。

7 标识

除符合《预包装食品标识通用标准》（CXS 1-1985）和《乳品术语》（CXS 206-1999）外，还应符合下列具体规定。

7.1 产品名称

卡门培尔干酪和卡雷德卡门培尔干酪名称应符合《预包装食品标识通用标准》（CXS 1-1985）4.1条，只要产品符合该标准，产品销售国的习惯拼写也可被使用。

"卡雷德"术语可被产品销售国适用的其他适当术语所取代。

只要干酪符合本标准，就可选择使用此名称。对符合本标准但不使用此名称的干酪，应按《干酪通则》（CXS 283-1978）命名规定来命名。

对于脂肪含量超过本标准第3.3条规定的参考含量范围，但高于绝对最低含量的产品，在命名时须附上适当修饰语，作为名称的一部分或处于与名称同一视野内的显眼位置，说明所做调整或脂肪含量（以干物质中脂肪含量或质量百分比表示，视产品销售国的接受情况）。适当修饰语是指《干酪通则》（CXS 283-1978）第7.2条所规定的特征描述用语或是符合《营养和保健声明使用准则》（CXG 23-1997）[1]的营养声明。

该名称也可用于符合本标准规定的干酪做成的切块、切片、切碎或磨碎产品。

7.2　原产国

应当标记原产国（即生产国，而非名称起源国）。当产品在第二个国家有实质性改造[2]，则以标识目的进行改造的国家应视为原产国。

7.3　乳脂含量说明

乳脂含量应以销售国可接受的方式声明：（i）质量百分比形式；（ii）干物质中脂肪百分比形式；（iii）如果产品标识上标明了份数，也可用每份中乳脂重量（g）表示。

7.4　非零售包装标识

在包装容器上除要标明产品名称、批次、生产厂家或包装商的名称和地址外，在包装容器上或附随说明书中，也应按本标准第7条和《预包装食品标识通用标准》（CXS 1-1985）第4.1～4.8条所要求的信息以及储存说明（在必要的地方）加以陈述。如果批次、生产厂家或包装商的名称和地址可以在附随的文件标明，则可以用一个识别标识来代替。

8　抽样和分析方法

为检查产品是否符合本标准，应采用《分析和抽样推荐性方法》（CXS 234-1999）中所涉及本标准规定的分析和抽样方法。

1　为比较营养声明，干物质中脂肪含量最低参考值为45%。
2　例如，重新包装、切块、切片、切碎和磨碎不构成实质性改变。

附录

附 加 信 息

以下补充信息对前文各条款的规定不构成影响,前文规定内容对产品标识、食品名称的使用以及食品安全性至关重要。

1　加工方法

1.1　发酵过程:微生物产酸发酵。

1.2　凝固类型:乳蛋白凝固通常是合适凝固温度下微生物酸化和蛋白酶(如凝乳酶)的共同作用结果。

STANDARD FOR CAMEMBERT

CXS 276-1973

Formerly CODEX STAN C-33-1973. Adopted in 1973. Revised in 2007.
Amended in 2008, 2010, 2018, 2019.

1 SCOPE

This Standard applies to Camembert intended for direct consumption or for further processing in conformity with the description in Section 2 of this Standard.

2 DESCRIPTION

Camembert is a soft surface ripened, primarily mould ripened cheese in conformity with the *General Standard for Cheese* (CXS 283-1978), which has a shape of a flat cylinder or sectors thereof. The body has a near white through to light yellow colour and a soft-textured (when pressed by thumb), but not crumbly texture, ripened from the surface to the center of the cheese. Gas holes are generally absent, but few openings and splits are acceptable. A rind is to be developed that is soft and entirely covered with white mould but may have red, brownish or orange coloured spots. Whole cheese may be cut or formed into sectors prior to or after the mould development.

For Camembert ready for consumption, the ripening procedure to develop flavour and body characteristics is normally from 10 days at 10–16 °C depending on the extent of maturity required. Alternative ripening conditions (including the addition of ripening enhancing enzymes) may be used, provided the cheese exhibits similar physical, biochemical and sensory properties as those achieved by the previously stated ripening procedure. Camembert intended for further processing need not exhibit the same extent of ripening when justified through technical and/or trade needs.

Carré de Camembert is a soft surface ripened cheese with a square shape and which comply with all other criteria and requirements specified for Camembert.

3 ESSENTIAL COMPOSITION AND QUALITY FACTORS

3.1 Raw materials

Cows' milk or buffaloes' milk, or their mixtures, and products obtained from these milks.

3.2 Permitted ingredients

- Starter cultures of harmless lactic acid and/or flavour producing bacteria and cultures of other harmless micro-organisms, including *Geotrichum candidum*, *Brevibacterium linens*, and yeast;
- Rennet or other safe and suitable coagulating enzymes;
- Sodium chloride and potassium chloride as a salt substitute;
- Potable water;
- Safe and suitable enzymes to enhance the ripening process;

- Safe and suitable processing aids;

- Rice, corn and potato flours and starches: Notwithstanding the provisions in the *General Standard for Cheese* (CXS 283-1978), these substances can be used in the same function as anti-caking agents for treatment of the surface of cut, sliced, and shredded products only, provided they are added only in amounts functionally necessary as governed by Good Manufacturing Practice, taking into account any use of the anti-caking agents listed in Section 4.

3.3 Composition

Milk constituent	Minimum content (m/m)	Maximum content (m/m)	Reference level (m/m)
Milkfat in dry matter	30%	Not restricted	45% to 55%
Dry matter	Depending on the fat in dry matter content, according to the table below		
	Fat in dry matter content (m/m)		**Corresponding minimum dry matter content** (m/m)
	Equal to or above 30% but less than 40%		38%
	Equal to or above 40% but less than 45%		41%
	Equal to or above 45% but less than 55%		43%
	Equal to or above 55%		48%

Compositional modifications beyond the minima and maxima specified above for milkfat and dry matter are not considered to be in compliance with Section 4.3.3 of the *General Standard for the Use of Dairy Terms* (CXS 206-1999).

3.4 Essential sizes and shapes

Maximum height: Approx. 5 cm;

Weight: Whole cheese of flat cylinder (Camembert)
or square (Carré de Camembert): approx. 80 g to 500 g.

3.5 Essential ripening procedure

Rind formation and maturation (proteolysis) from the surface to the centre is predominantly caused by *Penicillium candidium* and/or *Penicillium camembertii* and *Penicillium caseicolum*.

4 FOOD ADDITIVES

Only those additives classes indicated as justified in the table below may be used for the product categories specified. Colours used in accordance with Tables 1 and 2 of the General Standard for Food Additives (CXS 192-1995) in food category 01.6.2.1 (Ripened cheese, includes rind) and only certain acidity regulators in Table 3 are acceptable for use in foods conforming to this standard.

STANDARD FOR CAMEMBERT

Additive functional class	Justified use	
	Cheese mass	Surface/rind treatment
Colours	×(a)	–
Bleaching agents	–	–
Acids	–	–
Acidity regulators	×	–
Stabilizers	–	–
Thickeners	–	–
Emulsifiers	–	–
Antioxidants	–	–
Preservatives	–	–
Foaming agents	–	–
Anti–caking agents	–	–

(a) Only to obtain the colour characteristics, as described in Section 2.

× The use of additives belonging to the class is technologically justified.

– The use of additives belonging to the class is not technologically justified.

5 CONTAMINANTS

The products covered by this Standard shall comply with the maximum levels for contaminants that are specified for the product in the *General Standard for Contaminants and Toxins in Food and Feed* (CXS 193-1995).

The milk used in the manufacture of the products covered by this Standard shall comply with the maximum levels for contaminants and toxins specified for milk by the *General Standard for Contaminants and Toxins in Food and Feed* (CXS 193-1995) and with the maximum residue limits for veterinary drug residues and pesticides established for milk by the CAC.

6 HYGIENE

It is recommended that the product covered by the provisions of this Standard be prepared and handled in accordance with the appropriate sections of the *General Principles of Food Hygiene* (CXC 1-1969), the *Code of Hygienic Practice for Milk and Milk Products* (CXC 57-2004) and other relevant Codex texts such as Codes of Hygienic Practice and Codes of Practice. The products should comply with any microbiological criteria established in accordance with the *Principles and Guidelines for the Establishment and Application of Microbiological Criteria Related to Foods* (CXG 21-1997).

7 LABELLING

In addition to the provisions of the *General Standard for the Labelling of Prepackaged Foods* (CXS 1-1985) and the *General Standard for the Use of Dairy Terms* (CXS 206-

1999), the following specific provisions apply.

7.1 Name of the food

The names Camembert and Carré de Camembert may be applied in accordance with Section 4.1 of the *General Standard for the Labelling of Prepackaged Foods* (CXS 1-1985), provided that the product is in conformity with this Standard. Where customary in the country of retail sale, alternative spelling may be used.

The term "Carré de" may be replaced by other appropriate term(s) related to shape that are suitable in the country of retail sale.

The use of the names is an option that may be chosen only if the cheese complies with this Standard. Where the name is not used for a cheese that complies with this standard, the naming provisions of the *General Standard for Cheese* (CXS 283-1978) apply.

The designation of products in which the fat content is below or above the reference range but above the absolute minimum specified in Section 3.3 of this Standard shall be accompanied by an appropriate qualification describing the modification made or the fat content (expressed as fat in dry matter or as percentage by mass whichever is acceptable in the country of retail sale), either as part of the name or in a prominent position in the same field of vision. Suitable qualifiers are the appropriate characterizing terms specified in Section 7.2 of the *General Standard for Cheese* (CXS 283-1978) or a nutritional claim in accordance with the *Guidelines for Use of Nutrition and Health Claims* (CXG 23-1997)[1].

The designation may also be used for cut, sliced, shredded or grated products made from cheese which cheese is in conformity with this Standard.

7.2 Country of origin

The country of origin (which means the country of manufacture, not the country in which the name originated) shall be declared. When the product undergoes substantial transformation[2] in a second country, the country in which the transformation is performed shall be considered to be the country of origin for the purpose of labelling.

7.3 Declaration of milkfat content

The milk fat content shall be declared in a manner found acceptable in the country of retail sale. either (i) as a percentage by mass, (ii) as a percentage of fat in dry matter, or (iii) in grams per serving as quantified in the label, provided that the number of servings is stated.

7.4 Labelling of non retail containers

Information specified in Section 7 of this Standard and Sections 4.1 to 4.8 of the *General Standard for the Labelling of Prepackaged Foods* (CXS 1-1985) and, if necessary, storage instructions, shall be given either on the container or in accompanying documents, except that the name of the product, lot identification, and the name of the manufacturer or packer shall appear on the container, and in the absence of such a container, on the product itself. However, lot identification and the name and address may be replaced by

---

1 For the purpose of comparative nutritional claims, the minimum fat content of 45% fat in dry matter constitutes the reference.

2 For instance, repackaging, cutting, slicing, shredding and grating is not regarded as substantial transformation.

an identification mark, provided that such mark is clearly identifiable with the accompanying documents.

8 **METHODS OF SAMPLING AND ANALYSIS**

For checking the compliance with this Standard, the methods of analysis and sampling contained in the *Recommended Methods of Analysis and Sampling* (CXS 234-1999) relevant to the provisions in this Standard, shall be used.

APPENDIX

ADDITIONAL INFORMATION

The additional information below does not affect the provisions in the preceding sections which are those that are essential to the product identity, the use of the name of the food and the safety of the food.

1   Method of manufacture

1.1   Fermentation procedure: Microbiologically derived acid development.

1.2   Type of coagulation: Coagulation of the milk protein is typically obtained through the combined action of microbial acidification and proteases (e.g. rennet) at an appropriate coagulation temperature.

布 里 干 酪

STANDARD FOR BRIE

CXS 277-1973

前为CODEX STAN C-34-1973。1973年通过。2007年修订。
2008年、2010年、2018年、2019年修正。

1 范围

本标准适用于符合本标准第2条所述即食或用于进一步加工的布里干酪。

2 说明

布里干酪是一种软质表面成熟干酪，主要属于《干酪通则》（CXS 283-1978）中提及的霉菌成熟干酪，形状呈扁圆柱形或扇形。外观从近乎白色到浅黄色不等，质地柔软（用拇指摁压时），不易碎，从表面到中心完全成熟。布里干酪通常无气孔，但允许个别开口或裂口。干酪带有软质且完全被白霉覆盖的外皮，上面可有红色、褐色或橘色斑点。完整干酪在霉菌生成前后可切分或制成扇形。

对于即食的布里干酪，调制味道和外观特征的催熟程度在10～16 ℃下通常需要10天，取决于所需成熟度。在干酪能具备和前述催熟程序同等的物理、生化和感官特性的情况下，可以使用其他催熟条件（包括添加加速催熟的酶）。如有合理技术或贸易需求，用于进一步加工的布里干酪无需具备同等成熟度。

3 基本成分和质量要求

3.1 基本原料

牛乳、水牛乳或它们的混合乳、以及源于这些乳的制品。

3.2 其他配料

— 无害乳酸发酵剂和/或产香菌和其他无害微生物的培养剂；包括白地霉、亚麻短杆菌和酵母菌；

— 凝乳酶或其他安全、适宜的凝固酶；

— 作为代盐制品的氯化钠和氯化钾；

— 饮用水；

— 促进成熟的安全、适宜的酶；

— 安全、适宜的加工助剂；

— 大米、玉米、马铃薯粉和淀粉：尽管有《干酪通则》（CXS 283-1978），但这些物质同样可作为抗结块剂处理切块、切片和切碎产品表面，只要添加量是良好操作规范所要求达到期待功能的必需量，同时考虑第4条规定的抗结块剂的使用。

3.3 成分

乳成分	最低含量（m/m）	最高含量（m/m）	参考含量（m/m）
干物质乳脂	40%	/	45%~55%
干物质	根据干物质脂肪含量，参见下表		
	干物质脂肪含量（m/m）	相应干物质最低含量（m/m）	
	≥40%，<45%	42%	
	≥45%，<55%	43%	
	≥55%，<60%	48%	
	≥60%	51%	

乳脂和干物质含量超过规定的最大值和最小值被视为不符合《乳品术语》（CXS 206-1999）第4.3.3条。

3.4 基本尺寸和形状

最大高度：大约5 cm；

重量：扁圆柱形整干酪：约500~3 500 g。

3.5 基本熟化过程

外皮形成和从表面到中心的成熟过程（蛋白水解）主要在白青霉（*Penicillium candidium*）和/或卡门培尔青霉（*Penicillium camembertii*）和白酪青霉（*Penicillium caseicolum*）作用下促成。

4 食品添加剂

所述产品仅允许合理使用下表所列添加剂类别。在每个类别中，仅允许按照规定的功能和限量值使用表中所列添加剂。可以接受根据《食品添加剂通用标准》（CXS 192-1995）表1和表2用于食品类别01.6.2.1（成熟干酪，包括外皮）的着色剂，以及表3中仅部分酸度调节剂可用于符合本标准的食品。

添加剂功能类别	合理使用	
	酪体	表面/外皮处理
着色剂	×(a)	-
漂白剂	-	-
酸	-	-
酸度调节剂	×	-
稳定剂	-	-
增稠剂	-	-
乳化剂	-	-
抗氧化剂	-	-
防腐剂	-	-
发泡剂	-	-
抗结块剂	-	-

（a）仅为获得第2条所述颜色特征。

× 使用该类别添加剂符合技术要求。

- 使用该类别添加剂不符合技术要求。

5　污染物

本标准所涉及的产品应符合《食品及饲料中污染物和毒素通用标准》（CXS 193-1995）规定的污染物最大限量。

本标准所涉及的产品在加工中使用的牛乳应符合《食品及饲料中污染物和毒素通用标准》（CXS 193-1995）规定的乳中污染物和毒素最大限量以及国际食品法典委员会设定的农药和兽药最大残留限量。

6　卫生要求

建议本标准所涉及的产品应遵循《食品卫生总则》（CXC 1-1969）、《乳及乳制品卫生操作规范》（CXC 57-2004）以及《卫生操作规范》和《生产操作规范》等其他相关法典文本。产品应符合《食品微生物标准制定与实施原则和准则》（CXG 21-1997）规定的所有微生物标准。

7　标识

除符合《预包装食品标识通用标准》（CXS 1-1985）和《乳品术语》（CXS 206-

1999）外，还应符合以下具体规定。

7.1 产品名称

布里干酪名称应符合《预包装食品标识通用标准》（CXS 1-1985）第4.1条，只要产品符合该标准，产品销售国的习惯拼写也可被使用。

只要干酪符合本标准，就可选择使用此名称。对符合本标准但不使用此名称的干酪，应按《干酪通则》（CXS 283-1978）命名规定来命名。

对于脂肪含量超过本标准第3.3条规定的参考含量范围，但高于绝对最低含量的产品，在命名时须附上适当修饰语，作为名称的一部分或处于与名称同一视野内的显眼位置，说明所做调整或脂肪含量（以干物质中脂肪含量或质量百分比表示，视产品销售国的接受情况）。适当修饰语是指《干酪通则》（CXS 283-1978）第7.2条所规定的特征描述用语或是符合《营养和保健声明使用准则》（CXG 23-1997）[1]的营养声明。

该名称也可用于符合本标准规定的干酪做成的切块、切片、切碎或磨碎产品。

7.2 原产国

应当标记原产国（即生产国，而非名称起源国）。当产品在第二个国家有实质性改造[2]，则以标识目的进行改造的国家应视为原产国。

7.3 乳脂含量说明

乳脂含量应以销售国可接受的方式声明：（i）质量百分比形式；（ii）干物质中脂肪百分比形式；（iii）如果产品标识上标明了份数，也可用每份中乳脂重量（g）表示。

7.4 非零售包装标识

在包装容器上除要标明产品名称、批次、生产厂家或包装商的名称和地址外，在包装容器上或附随说明书中，也应按本标准第7条和《预包装食品标识通用标准》（CXS 1-1985）第4.1～4.8条所要求的信息以及储存说明（在必要的地方）加以陈述。如果批次、生产厂家或包装商的名称和地址可以在附随的文件标明，则可以用一个识别标识来代替。

8 抽样和分析方法

为检查产品是否符合本标准，应采用《分析和抽样推荐性方法》（CXS 234-1999）中涉及本标准规定的分析和抽样方法。

---

1 为比较营养声明，干物质中脂肪含量最低参考值为45%。
2 例如，重新包装、切块、切片、切碎和磨碎不构成实质性改变。

附录

附 加 信 息

以下补充信息对前文各条款的规定不构成影响,前文规定内容对产品标识、食品名称的使用以及食品安全性至关重要。

1 生产方法

1.1 发酵程序:微生物产酸发酵。

1.2 凝固类型:乳蛋白凝固通常是合适凝固温度下微生物酸化和蛋白酶(如凝乳酶)的共同作用结果。

STANDARD FOR BRIE

CXS 277-1973

Formerly CODEX STAN C-34-1973. Adopted in 1973. Revised in 2007.

Amended in 2008, 2010, 2018, 2019.

1 SCOPE

This Standard applies to Brie intended for direct consumption or for further processing in conformity with the description in Section 2 of this Standard.

2 DESCRIPTION

Brie is a soft surface ripened, primarily white mould ripened cheese in conformity with the *General Standard for Cheese* (CXS 283-1978), which has a shape of a flat cylinder or sectors thereof. The body has a near white through to light yellow colour and a soft-textured (when thumbs-pressed), but not crumbly texture, ripened from the surface to the center of the cheese. Gas holes are generally absent, but few openings and splits are acceptable. A rind is to be developed that is soft and entirely covered with white mould but may have red, brownish or orange coloured spots. Whole cheese may be cut or formed into sectors prior to or after the mould development.

For Brie ready for consumption, the ripening procedure to develop flavour and body characteristics is normally from 10 days at 10–16 °C depending on the extent of maturity required. Alternative ripening conditions (including the addition of ripening enhancing enzymes) may be used, provided the cheese exhibits similar physical, biochemical and sensory properties as those achieved by the previously stated ripening procedure. Brie intended for further processing need not exhibit the same extent of ripening when justified through technical and/or trade needs.

3 ESSENTIAL COMPOSITION AND QUALITY FACTORS

3.1 Raw materials

Cows' milk or buffaloes' milk, or their mixtures, and products obtained from these milks.

3.2 Permitted ingredients

- Starter cultures of harmless lactic acid and/or flavour producing bacteria and cultures of other harmless micro-organisms, including *Geotrichum candidum*, *Brevibacterium linens*, and yeast;
- Rennet or other safe and suitable coagulating enzymes;
- Sodium chloride and potassium chloride as a salt substitute;
- Potable water;
- Safe and suitable enzymes to enhance the ripening process;
- Safe and suitable processing aids;
- Rice, corn and potato flours and starches: Notwithstanding the provisions in the *General Standard for Cheese* (CXS 283-1978), these substances can be used in

the same function as anti-caking agents for treatment of the surface of cut, sliced, and shredded products only, provided they are added only in amounts functionally necessary as governed by Good Manufacturing Practice, taking into account any use of the anti-caking agents listed in Section 4.

3.3 Composition

Milk constituent	Minimum content (m/m)	Maximum content (m/m)	Reference level (m/m)
Milkfat in dry matter	40%	Not restricted	45% to 55%
Dry matter	Depending on the fat in dry matter content, according to the table below		
	Fat in dry matter content (m/m)	Corresponding minimum dry matter content (m/m)	
	Equal to or above 40% but less than 45%	42%	
	Equal to or above 45% but less than 55%	43%	
	Equal to or above 55% but less than 60%	48%	
	Equal to or above 60%	51%	

Compositional modifications beyond the minima and maxima specified above for milkfat and dry matter are not considered to be in compliance with Section 4.3.3 of the *General Standard for the Use of Dairy Terms* (CXS 206-1999).

3.4 Essential sizes and shapes

Maximum height: Approx. 5 cm;

Weight: Whole cheese of flat cylinder: approx. 500 g to 3 500 g.

3.5 Essential ripening procedure

Rind formation and maturation (proteolysis) from the surface to the centre is predominantly caused by *Penicillium candidium* and/or *Penicillium camembertii* and *Penicillium caseicolum*.

4 FOOD ADDITIVES

Only those additives classes indicated as justified in the table below may be used for the product categories specified. Colours used in accordance with Tables 1 and 2 of the *General Standard for Food Additives* (CXS 192-1995) in food category 01.6.2.1 (Ripened cheese, includes rind) and only certain acidity regulators in Table 3 are acceptable for use in foods conforming to this standard.

Additive functional class	Justified use	
	Cheese mass	Surface/rind treatment
Colours	x[a]	–
Bleaching agents	–	–
Acids	–	–

Additive functional class	Justified use	
	Cheese mass	Surface/rind treatment
Acidity regulators	×	–
Stabilizers	–	–
Thickeners	–	–
Emulsifiers	–	–
Antioxidants	–	–
Preservatives	–	–
Foaming agents	–	–
Anti–caking agents	–	–

(a) Only to obtain the colour characteristics, as described in Section 2.

× The use of additives belonging to the class is technologically justified.

– The use of additives belonging to the class is not technologically justified.

5 CONTAMINANTS

The products covered by this Standard shall comply with the maximum levels for contaminants that are specified for the product in the *General Standard for Contaminants and Toxins in Food and Feed* (CXS 193-1995).

The milk used in the manufacture of the products covered by this Standard shall comply with the maximum levels for contaminants and toxins specified for milk by the *General Standard for Contaminants and Toxins in Food and Feed* (CXS 193-1995) and with the maximum residue limits for veterinary drug residues and pesticides established for milk by the CAC.

6 HYGIENE

It is recommended that the product covered by the provisions of this Standard be prepared and handled in accordance with the appropriate sections of the *General Principles of Food Hygiene* (CXC 1-1969), the *Code of Hygienic Practice for Milk and Milk Products* (CXC 57-2004) and other relevant Codex texts such as Codes of Hygienic Practice and Codes of Practice The products should comply with any microbiological criteria established in accordance with the *Principles and Guidelines for the Establishment and Application of Microbiological Criteria Related to Foods* (CXG 21-1997).

7 LABELLING

In addition to the provisions of the *General Standard for the Labelling of Prepackaged Foods* (CXS 1-1985) and the *General Standard for the Use of Dairy Terms* (CXS 206-1999), the following specific provisions apply.

7.1 Name of the food

The name Brie may be applied in accordance with Section 4.1 of the *General Standard for the Labelling of Prepackaged Foods* (CXS 1-1985), provided that the product is in conformity with this Standard. Where customary in the country of retail sale, alternative spelling may be used.

The use of the name is an option that may be chosen only if the cheese complies with this Standard. Where the name is not used for a cheese that complies with this Standard, the naming provisions of the *General Standard for Cheese* (CXS 283-1978) apply.

The designation of products in which the fat content is below or above the reference range but above the absolute minimum specified in Section 3.3 of this Standard shall be accompanied by an appropriate qualification describing the modification made or the fat content (expressed as fat in dry matter or as percentage by mass whichever is acceptable in the country of retail sale), either as part of the name or in a prominent position in the same field of vision. Suitable qualifiers are the appropriate characterizing terms specified in Section 7.2 of the *General Standard for Cheese* (CXS 283-1978) or a nutritional claim in accordance with the *Guidelines for Use of Nutrition and Health Claims* (CXG 23-1997)[1].

The designation may also be used for cut, sliced, shredded or grated products made from cheese which cheese is in conformity with this Standard.

7.2 Country of origin

The country of origin (which means the country of manufacture, not the country in which the name originated) shall be declared. When the product undergoes substantial transformation[2] in a second country, the country in which the transformation is performed shall be considered to be the country of origin for the purpose of labelling.

7.3 Declaration of milkfat content

The milk fat content shall be declared in a manner found acceptable in the country of retail sale, either (i) as a percentage by mass, (ii) as a percentage of fat in dry matter, or (iii) in grams per serving as quantified in the label, provided that the number of servings is stated.

7.4 Labelling of non-retail containers

Information specified in Section 7 of this Standard and Sections 4.1 to 4.8 of the *General Standard for the Labelling of Prepackaged Foods* (CXS 1-1985) and, if necessary, storage instructions, shall be given either on the container or in accompanying documents, except that the name of the product, lot identification, and the name of the manufacturer or packer shall appear on the container, and in the absence of such a container, on the product itself. However, lot identification and the name and address may be replaced by an identification mark, provided that such mark is clearly identifiable with the accompanying documents.

8 METHODS OF SAMPLING AND ANALYSIS

For checking the compliance with this Standard, the methods of analysis and sampling contained in the *Recommended Methods of Analysis and Sampling* (CXS 234-1999) relevant to the provisions in this Standard, shall be used.

---

1 For the purpose of comparative nutritional claims, the minimum fat content of 45% fat in dry matter constitutes the reference.

2 For instance, repackaging, cutting, slicing, shredding and grating are not regarded as substantial transformation.

APPENDIX

ADDITIONAL INFORMATION

The additional information below does not affect the provisions in the preceding sections which are those that are essential to the product identity, the use of the name of the food and the safety of the food.

1 **Method of manufacture**

1.1 Fermentation procedure: Microbiologically derived acid development.

1.2 Type of coagulation: Coagulation of the milk protein is typically obtained through the combined action of microbial acidification and proteases (e.g. rennet) at an appropriate coagulation temperature.

特硬搓碎干酪

STANDARD FOR EXTRA HARD GRATING CHEESE

CXS 278-1978

原为CODEX STAN C-35-1978。1978年通过。2018年修正。

1　干酪名称

特硬搓碎干酪。

2　存放国

美国。

3　基本原料

3.1　乳品种：牛乳、山羊乳或绵羊乳，以及这些乳混合物。

3.2　允许添加物

3.2.1　必要添加物：

- 产生乳酸的无害细菌培养物（发酵剂）；
- 凝乳酶或其他适宜的凝固酶；
- 氯化钠。

3.2.2　可选添加物：

- 氯化钙，在所用牛乳中最大用量为无水氯化钙200 mg/kg；
- 产生香味的无害细菌；
- 有助于产生香味的无害酶剂（固体制剂不得超过所用乳重量0.1%）；
- 叶绿素，包括铜络复合物，在干酪中最大用量为15 mg/kg；
- 山梨酸、山梨酸钠盐或山梨酸钾盐，按山梨酸计算，在成品中最大用量为1 g/kg。

4　即食干酪主要特征

4.1　类型

4.1.1　一致性：特硬，适于搓碎。
4.1.2　简要说明：特硬、干燥、略脆、适于搓碎。食品加工期至少6个月。

4.2　形状：各种各样。

4.3　尺寸和重量：各种各样。

4.4 外皮（指有外皮干酪）

4.4.1 一致性：特硬。

4.4.2 外观：干燥，可具用植物油脂、食品级蜡或塑料材料的涂层。

4.4.3 色泽：琥珀色。

4.5 干酪体

4.5.1 质地：颗粒状，略脆。

4.5.2 色泽：自然无色至淡黄色。

4.6 气孔（气孔为该品种典型特征时）

4.6.1 数量：少量。

4.6.2 形状：小而圆。

4.6.3 大小规格：约1～2 mm。

4.6.4 外观：特征性的气孔。

4.7 **干物质中最低脂肪含量**：32%

4.8 **最大水分含量**：36%

5 加工方法

5.1 加工方法：凝乳酶或其他适宜凝固酶；可能添加乳酸发酵剂。

5.2 热处理：可用原乳或经巴氏法灭菌乳。如果用巴氏法灭菌，乳加热时温度不低于72 ℃（华氏161 ℉），处理15 s。

5.3 发酵过程：乳酸发酵或其他产生香味的培养物和酶。

5.4 熟化过程：凝乳可轻度腌制，然后模压成形，干酪可再次在盐水中腌制、干燥腌制或二者兼具；可置于通风或温度可控制的房间中，低温保存至少6个月。

6 抽样和分析方法

为检查是否符合本标准，应采用与本标准规定有关的《分析和抽样推荐性方法》（CXS 234-1999）中包含的分析和抽样方法。

7 标识及标签

7.1 只有符合本标准的干酪可用"特硬搓碎干酪"或在消费国家中公认的其他名称命名。如果不会产生误导且名称中带有"特硬搓碎干酪"词组可采用"新颖"或"奇特"的名称。

7.2 应按照《干酪通则》（CXS 283-1978）对产品加以标识。

STANDARD FOR EXTRA HARD GRATING CHEESE

CXS 278-1978

Formerly CODEX STAN C-35-1978. Adopted in 1978. Amended in 2018.

1 DESIGNATION OF CHEESE

Extra Hard Grating.

2 DEPOSITING COUNTRY

United States of America.

3 RAW MATERIALS

3.1 Kind of milk: Cow's milk, goat's milk or sheep's milk and mixtures of these milks.

3.2 Authorized additions

3.2.1 *Necessary additions:*

- cultures of harmless lactic acid producing bacteria (starter);
- rennet or other suitable coagulating enzymes;
- sodium chloride.

3.2.2 *Optional additions:*

- calcium chloride, max. 200 mg anhydrous/kg of the milk used;
- harmless flavour producing bacteria;
- harmless enzymes to assist in flavour development (solids of preparation not to exceed 0.1% of weight of milk used);
- chlorophyll, including copper chlorophyll complex, max. 15 mg/kg cheese;
- sorbic acid or its sodium or potassium salts, maximum 1 g/kg calculated as sorbic acid in the final product.

4 PRINCIPAL CHARACTERISTICS OF THE CHEESE READY FOR CONSUMPTION

4.1 Type

4.1.1 *Consistency:* Extra hard, suitable for grating.

4.1.2 *Short description:* Extra hard, dry, slightly brittle, suitable for grating. Period of curing at least 6 months.

4.2 Shape: Various.

4.3 Dimensions and weights: Various.

4.4 Rind (where present)

4.4.1 *Consistency:* Extra hard.

4.4.2 *Appearance:* Dry, may be coated with vegetable oil, food grade wax or plastic materials.

4.4.3 *Colour:* Amber.

4.5	**Body**
4.5.1	*Texture:* Granular, slightly brittle.
4.5.2	*Colour:* Natural uncoloured to light cream colour.
4.6	**Holes** (when holes are a typical characteristic of the variety)
4.6.1	*Number:* Few.
4.6.2	*Shape:* Small, round.
4.6.3	*Size:* Approximately 1–2 mm.
4.6.4	*Appearance:* Characteristic gas holes.
4.7	**Minimum fat content in dry matter:** 32%
4.8	**Maximum moisture content:** 36%

5 METHOD OF MANUFACTURE

5.1 **Method of coagulation:** Rennet or other suitable coagulating enzymes; with the possible addition of a lactic acid starter.

5.2 **Heat treatment:** Milk may be raw or pasteurized. If pasteurized the milk is heated to not less than 72 °C (161 °F) for 15 seconds.

5.3 **Fermentation procedure:** Lactic acid fermentation or other flavour producing cultures and enzymes.

5.4 **Maturation procedure:** After the curd which may be lightly salted is shaped into forms, the cheese may again be salted in brine, dry salted or both; held in a cool and well aerated or temperature controlled room for not less than 6 months.

6 METHODS OF SAMPLING AND ANALYSIS

For checking the compliance with this Standard, the methods of analysis and sampling contained in the *Recommended Methods of Analysis and Sampling* (CXS 234-1999) relevant to the provisions in this Standard, shall be used.

7 MARKING AND LABELLING

7.1 Only cheese conforming with this Standard may be designated Extra Hard Grating Cheese or any recognized variety name in the consuming country. A "coined" or "fanciful" name, may be used however, provided it is not misleading and is accompanied by the phrase "Extra Hard Grating Cheese".

7.2 It shall be labelled in conformity with the appropriate sections of the *General Standard for Cheese* (CXS 283-1978).

黄　油

STANDARD FOR BUTTER

CXS 279-1971

原为CODEX STAN A-1-1971。1971年通过。1999年修订。
2003年、2006年、2010年、2018年修正。

1　范围

本标准适用于符合本标准第2条所述即食或用于进一步加工的黄油。

2　说明

黄油是指完全来自乳和/或乳制品的脂肪产品，主要是以典型水油混合乳状液形式存在。

3　基本成分和质量要求

3.1　基本原料

乳和/或乳制品。

3.2　其他配料

- 氯化钠和食品级食盐；
- 无害乳酸发酵剂和/或产香菌；
- 饮用水。

3.3　成分

最低乳脂含量	80% m/m
最高水分含量	16% m/m
非脂乳固体最大含量	2% m/m

4　食品添加剂

根据本标准相关规定，可以采用《食品添加剂通用标准》（CXS 192-1995）中食品类别02.2.1（黄油）表1和表2中列出的食品添加剂。

5　污染物

本标准所涉及的产品应符合《食品及饲料中污染物和毒素通用标准》（CXS 193-1995）规定的污染物最大限量。

本标准所涉及的产品在加工中使用的牛乳应符合《食品及饲料中污染物和毒素通用标准》（CXS 193-1995）规定的乳中污染物和毒素最大限量以及国际食品法典委

会设定的农药和兽药最大残留限量。

6 卫生要求

建议本标准所涉及的产品应遵循《食品卫生总则》（CXC 1-1969）、《乳及乳制品卫生操作规范》（CXC 57-2004）以及《卫生操作规范》和《生产操作规范》等其他相关法典文本。产品应符合《食品微生物标准制定与实施原则和准则》（CXG 21-1997）规定的所有微生物标准。

7 标识

除符合《预包装食品标识通用标准》（CXS 1-1985）和《乳品术语》（CXS 206-1999）外，还应符合下列具体规定。

7.1 产品名称

产品名称应为"黄油"。"黄油"一词附带相应限定名称可以用作脂肪含量大于95%黄油的命名。

7.1.1 根据国家法规，可在黄油标签上标明其是否加盐。

7.2 乳脂含量说明

如果省略乳脂含量可能误导消费者，则应以产品最终消费国可以接受的方式声明：（i）质量百分比形式；（ii）干物质中脂肪百分比形式；（iii）如果产品标识上标明了份数，也可用每份中乳脂重量（g）表示。

7.3 非零售包装标识

在包装容器上除要标明产品名称、批次、生产厂家或包装商的名称和地址外，在包装容器上或附随说明书中，也应按本标准第7条和《预包装食品标识通用标准》（CXS 1-1985）第4.1~4.8条所要求的信息以及储存说明（在必要的地方）加以陈述。如果批次、生产厂家或包装商的名称和地址可以在附随的文件标明，则可以用一个识别标识来代替。

8 抽样和分析方法

为检查产品是否符合本标准，应采用《分析和抽样推荐性方法》（CXS 234-1999）中涉及本标准规定的分析和抽样方法。

STANDARD FOR BUTTER

CXS 279-1971

Formerly CODEX STAN A-1-1971. Adopted in 1971. Revised in 1999.

Amended in 2003, 2006, 2010, 2018.

1 SCOPE

This Standard applies to butter intended for direct consumption or for further processing in conformity with the description in Section 2 of this Standard.

2 DESCRIPTION

Butter is a fatty product derived exclusively from milk and/or products obtained from milk, principally in the form of an emulsion of the type water-in-oil.

3 ESSENTIAL COMPOSITION AND QUALITY FACTORS

3.1 Raw materials

Milk and/or products obtained from milk.

3.2 Permitted ingredients

- Sodium chloride and food grade salt;
- Starter cultures of harmless lactic acid and/or flavour producing bacteria;
- Potable water.

3.3 Composition

Minimum milkfat content	80% m/m
Maximum water content	16% m/m
Maximum milk solids-not-fat content	2% m/m

4 FOOD ADDITIVES

Food additives listed in Tables 1 and 2 of the *General Standard for Food Additives* (CXS 192-1995) in Food Category 02.2.1 (Butter) may be used in foods subject to this standard.

5 CONTAMINANTS

The products covered by this Standard shall comply with the maximum levels for contaminants that are specified for the product in the *General Standard for Contaminants and Toxins in Food and Feed* (CXS 193-1995).

The milk used in the manufacture of the products covered by this Standard shall comply with the maximum levels for contaminants and toxins specified for milk by the *General Standard for Contaminants and Toxins in Food and Feed* (CXS 193-1995) and with the maximum residue limits for veterinary drug residues and pesticides established for milk by the CAC.

6 HYGIENE

It is recommended that the products covered by the provisions of this standard be prepared and handled in accordance with the appropriate sections of the *General Principles of Food Hygiene* (CXC 1-1969), the *Code of Hygienic Practice for Milk and Milk Products* (CXC 57-2004) and other relevant Codex texts such as Codes of Hygienic Practice and Codes of Practice. The products should comply with any microbiological criteria established in accordance with the *Principles and Guidelines for the Establishment and Application of Microbiological Criteria Related to Foods* (CXG 21-1997).

7 LABELLING

In addition to the provisions of the General *Standard for the Labelling of Prepackaged Foods* (CXS 1-1985) and the *General Standard for the Use of Dairy Terms* (CXS 206-1999), the following specific provisions apply.

7.1 Name of the food

The name of the food shall be "Butter". The name "butter" with a suitable qualification shall be used for butter with more than 95% fat.

7.1.1
Butter may be labelled to indicate whether it is salted or unsalted according to national legislation.

7.2 Declaration of milkfat content

If the consumer would be misled by the omission, the milkfat content shall be declared in a manner found acceptable in the country of sale to the final consumer, either (i) as a percentage by mass, or (ii) in grams per serving as quantified in the label provided that the number of servings is stated.

7.3 Labelling of non-retail containers

Information required in Section 7 of this Standard and Sections 4.1 to 4.8 of the *General Standard for the Labelling of Prepackaged Foods* (CXS 1-1985), and, if necessary, storage instructions, shall be given either on the container or in accompanying documents, except that the name of the product, lot identification, and the name and address of the manufacturer or packer shall appear on the container. However, lot identification, and the name and address of the manufacturer or packer may be replaced by an identification mark, provided that such a mark is clearly identifiable with the accompanying documents.

8 METHODS OF SAMPLING AND ANALYSIS

For checking the compliance with this Standard, the methods of analysis and sampling contained in the *Recommended Methods of Analysis and Sampling* (CXS 234-1999) relevant to the provisions in this Standard, shall be used.

乳脂产品

STANDARD FOR MILKFAT PRODUCTS

CXS 280-1973

原为CODEX STAN A-2-1973。1973年通过。1999年修订。
2006年、2010年、2018年修正。

1　范围

本标准适用于符合本标准第2条所述用于进一步加工或用于烹饪的无水乳脂、乳脂、无水黄油、黄油和酥油。

2　说明

2.1　无水乳脂、乳脂、无水黄油和黄油是指完全来自乳和/或乳制品的脂肪产品，其通过去除几乎所有水分和非脂肪固体制成。

2.2　酥油是指完全来自乳、稀奶油或黄油并具有特殊香味和物理结构的产品，其通过去除几乎所有水分和非脂肪固体制成。

3　基本成分和质量要求

3.1　基本原料

乳和/或乳制品。

3.2　其他配料

无害乳酸发酵剂。

3.3　成分

成分	无水乳脂/无水黄油	乳脂	黄油	酥油
乳脂最低含量（% m/m）	99.8	99.6	99.6	99.6
水分最高含量（% m/m）	0.1	—	—	—

4　食品添加剂

根据本标准相关规定，宜使用《食品添加剂通用标准》（CXS 192-1995）表1和表2中食品类别02.1.1（黄油、无水乳脂、酥油）所列食品添加剂。

4.1　在装入产品前、中和后的密闭容器中所灌充的惰性气体。

5 污染物

本标准所涉及的产品应符合《食品及饲料中污染物和毒素通用标准》（CXS 193-1995）规定的污染物最大限量。

本标准所涉及的产品在加工中使用的牛乳应符合《食品及饲料中污染物和毒素通用标准》（CXS 193-1995）规定的乳中污染物和毒素最大限量以及国际食品法典委员会设定的农药和兽药最大残留限量。

6 卫生要求

建议本标准所涉及的产品应遵循《食品卫生总则》（CXC 1-1969）、《乳及乳制品卫生操作规范》（CXC 57-2004）以及《卫生操作规范》和《生产操作规范》等其他相关法典文本。产品应符合《食品微生物标准制定与实施原则和准则》（CXG 21-1997）规定的所有微生物标准。

7 标识

除符合《预包装食品标识通用标准》（CXS 1-1985）和《乳品术语》（CXS 206-1999）外，还应符合下列具体规定。

7.1 产品名称

产品名称应为：

无水乳脂	
乳脂	
无水黄油	根据第2条说明、第3条规定成分及抗氧化剂使用（见第4条）命名
黄油	
酥油	

7.2 非零售包装标识

在包装容器上除要标明产品名称、批次、生产厂家或包装商的名称和地址外，在包装容器上或附随说明书中，也应按本标准第7条和《预包装食品标识通用标准》（CXS 1-1985）第4.1～4.8条所要求的信息以及储存说明（在必要的地方）加以陈述。如果批次、生产厂家或包装商的名称和地址可以在附随的文件标明，则可以用一个识别标识来代替。

8 抽样和分析方法

为检查产品是否符合本标准，应采用《分析和抽样推荐性方法》（CXS 234-1999）中涉及本标准规定的分析和抽样方法。

附录

附 加 信 息

以下补充信息对前文各条款的规定不构成影响，前文规定内容对产品标识、食品名称的使用以及食品安全性至关重要。

1   **其他质量要求**

	无水乳脂/无水黄油	乳脂	黄油	酥油
游离脂肪酸最大含量（按油酸质量百分比计）	0.3	0.4	0.4	0.4
最大过氧化值（毫克当量氧/千克脂肪）	0.3	0.6	0.6	0.6
味道及气味	样品加热至40~45 ℃之后，达到市场上可接受的要求			
质地	光滑、细腻的液体，依温度而定			

2   **其他污染物**

重金属

以下限量适用于无水乳脂、乳脂、无水黄油、黄油和酥油：

金属	最大限量
铜	0.05 mg/kg
铁	0.2 mg/kg

3   **其他分析方法**

见CXS 234-1999。

STANDARD FOR MILKFAT PRODUCTS

CXS 280-1973

Formerly CODEX STAN A-2-1973. Adopted in 1973. Revised in 1999.

Amended in 2006, 2010, 2018.

1 SCOPE

This Standard applies to Anhydrous Milkfat, Milkfat, Anhydrous Butter oil, Butter oil and Ghee, which are intended for further processing or culinary use, in conformity with the description in Section 2 of this Standard.

2 DESCRIPTION

2.1 ***Anhydrous Milkfat, Milkfat, Anhydrous Butter oil and Butter oil*** are fatty products derived exclusively from milk and/or products obtained from milk by means of processes which result in almost total removal of water and non-fat solids.

2.2 ***Ghee*** is a product exclusively obtained from milk, cream or butter, by means of processes which result in almost total removal of water and non-fat solids, with an especially developed flavour and physical structure.

3 ESSENTIAL COMPOSITION AND QUALITY FACTORS

3.1 Raw materials

Milk and/or products obtained from milk.

3.2 Permitted ingredients

Starter cultures of harmless lactic acid producing bacteria.

3.3 Composition

	Anhydrous milkfat/ Anhydrous butter oil	Milkfat	Butter oil	Ghee
Minimum milkfat (% m/m)	99.8	99.6	99.6	99.6
Maximum water (% m/m)	0.1	–	–	–

4 FOOD ADDITIVES

Food additives listed in Tables 1 and 2 of the *General Standard for Food Additives* (CXS 192-1995) in Food Category 02.1.1 (Butter oil, anhydrous milkfat, ghee) may be used in foods subject to this Standard.

4.1 Inert gas with which airtight containers are flushed before, during and after filling with product.

5 CONTAMINANTS

The products covered by this Standard shall comply with the maximum levels for contaminants that are specified for the product in the *General Standard for Contaminants and Toxins in Food and Feed* (CXS 193-1995).

The milk used in the manufacture of the products covered by this Standard shall comply with the maximum levels for contaminants and toxins specified for milk by the *General Standard for Contaminants and Toxins in Food and Feed* (CXS 193-1995) and with the maximum residue limits for veterinary drug residues and pesticides established for milk by the CAC.

6 HYGIENE

It is recommended that the products covered by the provisions of this Standard be prepared and handled in accordance with the appropriate sections of the *General Principles of Food Hygiene* (CXC 1-1969), the *Code of Hygienic Practice for Milk and Milk Products* (CXC 57-2004) and other relevant Codex texts such as Codes of Hygienic Practice and Codes of Practice. The products should comply with any microbiological criteria established in accordance with the *Principles and Guidelines for the Establishment and Application of Microbiological Criteria Related to Foods* (CXG 21-1997).

7 LABELLING

In addition to the provisions of the *General Standard for the Labelling of Prepackaged Foods* (CXS 1-1985) and the *General Standard for the Use of Dairy Terms* (CXS 206-1999), the following specific provisions apply.

7.1 Name of the food

The name of the food shall be:

Anhydrous milkfat	
Milkfat	
Anhydrous butter oil	According to description specified in Section 2, composition specified in 3 and the use of antioxidants (see Section 4)
Butter oil	
Ghee	

7.2 Labelling of non-retail containers

Information required in Section 7 of this Standard and Sections 4.1 to 4.8 of the *General Standard for the Labelling of Prepackaged Foods* (CXS 1-1985), and, if necessary, storage instructions, shall be given either on the container or in accompanying documents, except that the name of the product, lot identification, and the name and address of the manufacturer or packer shall appear on the container. However, lot identification, and the name and address of the manufacturer or packer may be replaced by an identification mark, provided that such a mark is clearly identifiable with the accompanying documents.

8 METHODS OF SAMPLING AND ANALYSIS

For checking the compliance with this standard, the methods of analysis and sampling contained in the *Recommended Methods of Analysis and Sampling* (CXS 234-1999) relevant to the provisions in this standard, shall be used.

APPENDIX

ADDITIONAL INFORMATION

The additional information below does not affect the provisions in the preceding sections which are those that are essential to the product identity, the use of the name of the food and the safety of the food.

1 OTHER QUALITY FACTORS

	Anhydrous milkfat/ Anhydrous butter oil	Milkfat	Butter oil	Ghee
Maximum free fatty acids(% m/m as oleic acid)	0.3	0.4	0.4	0.4
Maximum peroxide value(milli-equivalents of oxygen/kg fat)	0.3	0.6	0.6	0.6
Taste and odour	Acceptable for market requirements after heating a sample to 40–45 °C			
Texture	Smooth and fine granules to liquid, depending on temperature			

2 OTHER CONTAMINANTS

Heavy metals

The following limits apply to Anhydrous Milkfat, Milkfat, Anhydrous Butter oil and Butter oil and Ghee:

Metal	Maximum level
Copper	0.05 mg/kg
Iron	0.2 mg/kg

3 OTHER METHODS OF ANALYSIS

See CXS 234-1999.

炼 乳

STANDARD FOR EVAPORATED MILKS

CXS 281-1971

1971年通过。1999年修订。2010年、2018年修正。

1 范围

本标准适用于符合本标准第2条所述即食或用于进一步加工的炼乳。

2 说明

炼乳是指通过加热或其他使产品保持原有成分和特性的加工方法，使乳部分脱水而制成的乳制品。可以调节乳脂和/或乳蛋白质含量，但仅调节符合本标准第3条成分要求的；为此添加和/或分离成分的同时，保证所调节乳中乳清蛋白与酪蛋白的比例不变。

3 基本成分和质量要求

3.1 基本原料

乳和乳粉[1]、稀奶油和稀奶油粉[1]、乳脂制品[1]。

允许使用下列乳制品调节蛋白含量：

- 乳超滤滞留物：乳超滤滞留物是指乳、部分脱脂乳或脱脂乳，通过超滤法浓缩乳蛋白后形成的产品；
- 乳渗透物：乳渗透物是指通过超滤法，将乳蛋白和乳脂从乳、部分脱脂乳或脱脂乳中分离除去而形成的产品；
- 乳糖[1]。

3.2 其他配料

- 饮用水；
- 氯化钠。

3.3 成分

炼乳

乳脂最低含量	7.5% m/m
乳固体最低含量[a]	25% m/m
非脂乳固体中乳蛋白最低含量[a]	34% m/m

---

1 见《食糖标准》（CXS 212-1999）。

脱脂炼乳

乳脂最高含量	1% m/m
乳固体最低含量 (a)	20% m/m
非脂乳固体中乳蛋白最低含量 (a)	34% m/m

部分脱脂炼乳

乳脂含量	>1%，<7.5% m/m
乳固体最低含量 (a)	20% m/m
非脂乳固体中乳蛋白最低含量 (a)	34% m/m

高脂炼乳

乳脂最低含量	15% m/m
非脂乳固体最低含量 (a)	11.5% m/m
非脂乳固体中乳蛋白最低含量 (a)	34% m/m

（a）乳固体和非脂乳固体含量包括乳糖结晶水。

4 食品添加剂

只允许在规定范围内使用下表列出的食品添加剂。

INS编号	添加剂名称	最大限量
固化剂		
508	氯化钾	2 000 mg/kg（单独使用）或3 000 mg/kg（混合使用），按无水物质计
509	氯化钙	
稳定剂		
331	柠檬酸钠	2 000 mg/kg（单独使用）或3 000 mg/kg（混合使用），按无水物质计
332	柠檬酸钾	
333	柠檬酸钙	
酸度调节剂		
170	碳酸钙	2 000 mg/kg（单独使用）或3 000 mg/kg（混合使用），按无水物质计
339	磷酸钠	
340	磷酸钾	
341	磷酸钙	
450	二磷酸盐	

INS编号	添加剂名称	最大限量
451	三磷酸盐	2 000 mg/kg（单独使用）或 3 000 mg/kg（混合使用），按无水物质计
452	聚磷酸盐	
500	碳酸钠	
501	碳酸钾	
增稠剂		
407	卡拉胶	150 mg/kg
乳化剂		
322	卵磷脂	根据GMP限量使用

5　污染物

本标准所涉及的产品应符合《食品及饲料中污染物和毒素通用标准》（CXS 193-1995）规定的污染物最大限量。

本标准所涉及的产品在加工中使用的牛乳应符合《食品及饲料中污染物和毒素通用标准》（CXS 193-1995）规定的乳中污染物和毒素最大限量以及国际食品法典委员会设定的农药和兽药最大残留限量。

6　卫生要求

建议本标准所涉及的产品应遵循《食品卫生总则》（CXC 1-1969）、《乳及乳制品卫生操作规范》（CXC 57-2004）以及《卫生操作规范》和《生产操作规范》等其他相关法典文本。产品应符合《食品微生物标准制定与实施原则和准则》（CXG 21-1997）规定的所有微生物标准。

7　标识

除符合《预包装食品标识通用标准》（CXS 1-1985）和《乳品术语》（CXS 206-1999）外，还应符合下列具体规定。

7.1　产品名称

产品名称应为：

炼乳	
脱脂炼乳	
部分脱脂炼乳	应根据第3条规定的成分命名
高脂炼乳	

如果部分脱脂炼乳乳脂含量为4.0%~4.5%，且乳固体最低含量为24% m/m，部分脱脂炼乳也可称为"半脱脂炼乳"。

7.2 乳脂含量说明

如果省略乳脂含量可能误导消费者，则应以产品最终消费国可以接受的方式声明：（i）质量百分比或体积百分比形式；（ii）如果产品标识上标明了份数，也可用每份中乳脂重量（g）表示。

7.3 乳蛋白含量说明

如果省略乳蛋白含量可能误导消费者，则应以产品最终消费国可接受方式标明：（i）质量百分比或体积百分比形式；（ii）如果产品标识上标明了份数，也可用每份中乳蛋白重量（g）表示。

7.4 配料表

尽管《预包装食品标识通用标准》（CXS 1-1985）第4.2.1条有规定，但仅用于调节蛋白含量的乳制品无需声明。

7.5 非零售包装标识

在包装容器上除要标明产品名称、批次、生产厂家或包装商的名称和地址外，在包装容器上或附随说明书中，也应按本标准第7条和《预包装食品标识通用标准》（CXS 1-1985）第4.1~4.8条所要求的信息以及储存说明（在必要的地方）加以陈述。如果批次、生产厂家或包装商的名称和地址可以在附随的文件标明，则可以用一个识别标识来代替。

8 抽样和分析方法

为检查产品是否符合本标准，应采用《分析和抽样推荐性方法》（CXS 234-1999）中涉及本标准规定的分析和抽样方法。

STANDARD FOR EVAPORATED MILKS

CXS 281-1971

Adopted in 1971. Revised in 1999. Amended in 2010, 2018.

1 SCOPE

This Standard applies to evaporated milks, intended for direct consumption or further processing, in conformity with the description in Section 2 of this Standard.

2 DESCRIPTION

Evaporated milks are milk products which can be obtained by the partial removal of water from milk by heat, or by any other process which leads to a product of the same composition and characteristics. The fat and/or protein content of the milk may have been adjusted, only to comply with the compositional requirements in Section 3 of this Standard, by the addition and/or withdrawal of milk constituents in such a way as not to alter the whey protein to casein ratio of the milk being adjusted.

3 ESSENTIAL COMPOSITION AND QUALITY FACTORS

3.1 Raw materials

Milk and milk powders[1], cream and cream powders[1], milkfat products[1].

The following milk products are allowed for protein adjustment purposes:

- Milk retentate: Milk retentate is the product obtained by concentrating milk protein by ultrafiltration of milk, partly skimmed milk, or skimmed milk;
- Milk permeate: Milk permeate is the product obtained by removing milk proteins and milkfat from milk, partly skimmed milk, or skimmed milk by ultrafiltration; and
- Lactose[1].

3.2 Permitted ingredients

- Potable water;
- Sodium chloride.

3.3 Composition

Evaporated milk

Minimum milkfat	7.5% m/m
Minimum milk solids[a]	25% m/m
Minimum milk protein in milk solids-not-fat[a]	34% m/m

Evaporated skimmed milk

Maximum milkfat	1% m/m

---

1 See *Standard for Sugars* (CXS 212-1999).

Minimum milk solids[a]	20% m/m
Minimum milk protein in milk solids-not-fat[a]	34% m/m

Evaporated partly skimmed milk

Milkfat	More than 1% and less than 7.5% m/m
Minimum milk solids[a]	20% m/m
Minimum milk protein in milk solids-not-fat[a]	34% m/m

Evaporated high-fat milk

Minimum milkfat	15% m/m
Minimum milk solids-not-fat[a]	11.5% m/m
Minimum milk protein in milk solids-not-fat[a]	34% m/m

(a) The milk solids and milk solids-not-fat content includes water of crystallization of the lactose.

4   FOOD ADDITIVES

Only those food additives listed below may be used and only within the limits specified.

INS no.	Name of additive	Maximum level
Firming agents		
508	Potassium chloride	2 000 mg/kg singly or 3 000 mg/kg in combination, expressed as anhydrous substances
509	Calcium chloride	
Stabilizers		
331	Sodium citrates	2 000 mg/kg singly or 3 000 mg/kg in combination, expressed as anhydrous substances
332	Potassium citrates	
333	Calcium citrates	
Acidity regulators		
170	Calcium carbonates	2 000 mg/kg singly or 3 000 mg/kg in combination, expressed as anhydrous substances
339	Sodium phosphates	
340	Potassium phosphates	
341	Calcium phosphates	
450	Diphosphates	
451	Triphosphates	
452	Polyphosphates	
500	Sodium carbonates	
501	Potassium carbonates	

INS no.	Name of additive	Maximum level
Thickener		
407	Carrageenan	150 mg/kg
Emulsifier		
322	Lecithins	Limited by GMP

5 CONTAMINANTS

The products covered by this Standard shall comply with the maximum levels for contaminants that are specified for the product in the *General Standard for Contaminants and Toxins in Food and Feed* (CXS 193-1995).

The milk used in the manufacture of the products covered by this Standard shall comply with the maximum levels for contaminants and toxins specified for milk by the *General Standard for Contaminants and Toxins in Food and Feed* (CXS 193-1995) and with the maximum residue limits for veterinary drug residues and pesticides established for milk by the CAC.

6 HYGIENE

It is recommended that the products covered by the provisions of this standard be prepared and handled in accordance with the appropriate sections of the *General Principles of Food Hygiene* (CXC 1-1969), the *Code of Hygienic Practice for Milk and Milk Products* (CXC 57-2004) and other relevant Codex texts such as Codes of Hygienic Practice and Codes of Practice. The products should comply with any microbiological criteria established in accordance with the *Principles and Guidelines for the Establishment and Application of Microbiological Criteria Related to Foods* (CXG 21-1997).

7 LABELLING

In addition to the provisions of the *General Standard for the Labelling of Prepackaged Foods* (CXS 1-1985) and the *General Standard for the Use of Dairy Terms* (CXS 206-1999), the following specific provisions apply.

7.1 Name of the food

The name of the food shall be:

Evaporated milk	
Evaporated skimmed milk	According to the composition specified in Section 3
Evaporated partly skimmed milk	
Evaporated high-fat milk	

Evaporated partly skimmed milk may be designated "evaporated semi-skimmed milk" if the milkfat content is 4.0%–4.5% and the minimum milk solids is 24% m/m.

7.2 Declaration of milkfat content

If the consumer would be misled by the omission, the milkfat content shall be declared in a manner found acceptable in the country of sale to the final consumer, either (i) as a percentage by mass or volume, or (ii) in grams per serving as quantified in the label provided that the number of servings is stated.

7.3　Declaration of milk protein

If the consumer would be misled by the omission, the milk protein content shall be declared in a manner acceptable in the country of sale to the final consumer, either as (i) a percentage by mass or volume, or (ii) grams per serving as quantified in the label provided that the number of servings is stated.

7.4　List of ingredients

Notwithstanding the provision of Section 4.2.1 of the *General Standard for the Labelling of Prepackaged Foods* (CXS 1-1985), milk products used only for protein adjustment need not be declared.

7.5　Labelling of non-retail containers

Information required in Section 7 of this Standards and Sections 4.1 to 4.8 of the *General Standard for the Labelling of Prepackaged Foods* (CXS 1-1985), and, if necessary, storage instructions, shall be given either on the container or in accompanying documents, except that the name of the product, lot identification, and the name and address of the manufacturer or packer shall appear on the container. However, lot identification, and the name and address of the manufacturer or packer may be replaced by an identification mark, provided that such a mark is clearly identifiable with the accompanying documents.

8　METHODS OF SAMPLING AND ANALYSIS

For checking the compliance with this standard, the methods of analysis and sampling contained in the *Recommended Methods of Analysis and Sampling* (CXS 234-1999) relevant to the provisions in this standard, shall be used.

甜 炼 乳

STANDARD FOR SWEETENED CONDENSED MILKS

CXS 282-1971

1971年通过。1999年修订。2010年、2018年修正。

1 范围

本标准适用于符合本标准第2条所述即食或用于进一步加工的甜炼乳。

2 说明

甜炼乳是指乳中加糖部分分离水分而成的乳制品，或是采用任何其他加工方法制成的成分和特点相同的产品。可以调节乳脂和/或乳蛋白含量，但仅调节符合本标准第3条成分要求，为此添加和/或分离乳成分的同时，保证所调节乳中乳清蛋白与酪蛋白比例不变。

3 基本成分和质量要求

3.1 基本原料

乳和乳粉[1]、稀奶油和稀奶油粉[1]、乳脂制品[1]。

允许使用下列乳制品调节蛋白质含量：

- 乳超滤滞留物：乳超滤滞留物是指乳、部分脱脂乳或脱脂乳，通过超滤法浓缩乳蛋白后形成的产品；
- 乳渗透物：乳渗透物是指通过超滤法，将乳蛋白和乳脂从乳、部分脱脂乳或脱脂乳中分离除去而形成的产品；
- 乳糖[1]。

3.2 其他配料

- 饮用水；
- 糖类；
- 氯化钠。

在本产品中，糖类通常考虑采用蔗糖，但可按照良好生产规范，将蔗糖与其他糖类混合。

3.3 成分

甜炼乳

---

1 见《糖类标准》（CXS 212-1999）。

乳脂最低含量	8% m/m
乳固体最低含量[a]	28% m/m
非脂乳固体中乳蛋白最低含量[a]	34% m/m

脱脂甜炼乳

乳脂最高含量	1% m/m
乳固体最低含量[a]	24% m/m
非脂乳固体中乳蛋白最低含量[a]	34% m/m

部分脱脂甜炼乳

乳脂含量	>1%, <8% m/m
非脂乳固体最低含量[a]	20% m/m
乳固体最低含量[a]	24% m/m
非脂乳固体中乳蛋白最低含量[a]	34% m/m

高脂甜炼乳

乳脂最低含量	16% m/m
非脂乳固体最低含量[a]	14% m/m
非脂乳固体中乳蛋白最低含量[a]	34% m/m

（a）乳固体和非脂乳固体含量包括乳糖结晶水。

对于所有甜炼乳，含糖量以良好生产规范为限，最低含量应保障产品耐储性，最高含量应避免可能出现糖类结晶。

4　食品添加剂

只允许在规定范围内使用下表列出的食品添加剂。

INS编号	添加剂名称	最大限量
固化剂		
508	氯化钾	2 000 mg/kg（单独使用）或3 000 mg/kg（混合使用），以无水物质计
509	氯化钙	
稳定剂		
331	柠檬酸钠	2 000 mg/kg（单独使用）或3 000 mg/kg（混合使用），以无水物质计
332	柠檬酸钾	
333	柠檬酸钙	

INS编号	添加剂名称	最大限量
酸度调节剂		
170	碳酸钙	2 000 mg/kg（单独使用）或 3 000 mg/kg（混合使用），以无水物质计
339	磷酸钠	
340	磷酸钾	
341	磷酸钙	
450	二磷酸盐	
451	三磷酸盐	
452	聚磷酸盐	
500	碳酸钠	
501	碳酸钾	
增稠剂		
407	卡拉胶	150 mg/kg
乳化剂		
322	卵磷脂	根据GMP限量使用

5 污染物

本标准所涉及的产品应符合《食品及饲料中污染物和毒素通用标准》（CXS 193-1995）规定的污染物最大限量。

本标准所涉及的产品在加工中使用的牛乳应符合《食品及饲料中污染物和毒素通用标准》（CXS 193-1995）规定的乳中污染物和毒素最大限量以及国际食品法典委员会设定的农药和兽药最大残留限量。

6 卫生要求

建议本标准所涉及的产品应遵循《食品卫生总则》（CXC 1-1969）、《乳及乳制品卫生操作规范》（CXC 57-2004）以及《卫生操作规范》和《生产操作规范》等其他相关法典文本。产品应符合《食品微生物标准制定与实施原则和准则》（CXG 21-1997）规定的所有微生物标准。

7 标识

除符合《预包装食品标识通用标准》（CXS 1-1985）和《乳品术语》（CXS 206-1999）外，还应符合以下具体规定。

7.1 产品名称

产品名称应为：

甜炼乳	
脱脂甜炼乳	参照第3条规定的成分
部分脱脂甜炼乳	
高脂甜炼乳	

如乳脂含量在4.0%~4.5%，并且乳固体最低含量为28% m/m，则部分脱脂炼乳可以标注为"半脱脂甜炼乳"。

7.2 乳脂含量说明

如果省略乳脂含量可能误导消费者，则应以产品最终消费国可以接受的方式声明：（i）质量百分比或体积百分比形式；（ii）如果产品标识上标明了份数，也可用每份中乳脂重量（g）表示。

7.3 乳蛋白含量说明

如果省略乳蛋白含量可能误导消费者，则应以产品最终消费国可接受方式标明：（i）质量百分比或体积百分比形式；（ii）如果产品标识上标明了份数，也可用每份中乳蛋白重量（g）表示。

7.4 配料表

尽管《预包装食品标识通用标准》（CXS 1-1985）第4.2.1条有规定，但仅用于调整乳蛋白含量的乳制品无需声明。

7.5 非零售包装标识

在包装容器上除要标明产品名称、批次、生产厂家或包装商的名称和地址外，在包装容器上或附随说明书中，也应按本标准第7条和《预包装食品标识通用标准》（CXS 1-1985）第4.1~4.8条所要求的信息以及储存说明（在必要的地方）加以陈述。如果批次、生产厂家或包装商的名称和地址可以在附随的文件标明，则可以用一个识别标识来代替。

8 抽样和分析方法

为检查产品是否符合本标准，应采用《分析和抽样推荐性方法》（CXS 234-1999）中涉及本标准规定的分析和抽样方法。

STANDARD FOR SWEETENED CONDENSED MILKS

CXS 282-1971

Adopted in 1971. Revised in 1999. Amended in 2010, 2018.

1 SCOPE

This Standard applies to sweetened condensed milks, intended for direct consumption or further processing, in conformity with the description in Section 2 of this Standard.

2 DESCRIPTION

Sweetened condensed milks are milk products which can be obtained by the partial removal of water from milk with the addition of sugar, or by any other process which leads to a product of the same composition and characteristics. The fat and/or protein content of the milk may have been adjusted, only to comply with the compositional requirements in Section 3 of this Standard, by the addition and/or withdrawal of milk constituents in such a way as not to alter the whey protein to casein ratio of the milk being adjusted.

3 ESSENTIAL COMPOSITION AND QUALITY FACTORS

3.1 Raw materials

Milk and milk powders[1], cream and cream powders[1], milkfat products[1].

The following milk products are allowed for protein adjustment purposes:

- Milk retentate: Milk retentate is the product obtained by concentrating milk protein by ultrafiltration of milk, partly skimmed milk, or skimmed milk;
- Milk permeate: Milk permeate is the product obtained by removing milk proteins and milkfat from milk, partly skimmed milk, or skimmed milk by ultrafiltration; and
- Lactose[1].

3.2 Permitted ingredients

- Potable water
- Sugar
- Sodium chloride.

In this product, sugar is generally considered to be sucrose, but a combination of sucrose with other sugars, consistent with Good Manufacturing Practice, may be used.

3.3 Composition

Sweetened condensed milk

Minimum milkfat	8% m/m
Minimum milk solids[a]	28% m/m
Minimum milk protein in milk solids-not-fat[a]	34% m/m

---

[1] See *Standard for Sugars* (CXS 212-1999).

Sweetened condensed skimmed milk

Maximum milkfat	1% m/m
Minimum milk solids[a]	24% m/m
Minimum milk protein in milk solids-not-fat[a]	34% m/m

Sweetened condensed partly skimmed milk

Milkfat	More than 1% and less than 8% m/m
Minimum milk solids-not-fat[a]	20% m/m
Minimum milk solids[a]	24% m/m
Minimum milk protein in milk solids-not-fat[a]	34% m/m

Sweetened condensed high-fat milk

Minimum milkfat	16% m/m
Minimum milk solids-not-fat[a]	14% m/m
Minimum milk protein in milk solids-not-fat[a]	34% m/m

(a) The milk solids and milk solids-not-fat content includes water of crystallization of the lactose.

For all sweetened condensed milks the amount of sugar is restricted by Good Manufacturing Practice to a minimum value which safeguards the keeping quality of the product and a maximum value above which crystallization of sugar, may occur.

4   FOOD ADDITIVES

Only those food additives listed below may be used and only within the limits specified.

INS no.	Name of additive	Maximum level
Firming agents		
508	Potassium chloride	2 000 mg/kg singly or 3 000 mg/kg in combination, expressed as anhydrous substances
509	Calcium chloride	
Stabilizers		
331	Sodium citrates	2 000 mg/kg singly or 3 000 mg/kg in combination, expressed as anhydrous substances
332	Potassium citrates	
333	Calcium citrates	
Acidity regulators		
170	Calcium carbonates	2 000 mg/kg singly or 3 000 mg/kg in combination, expressed as anhydrous substances
339	Sodium phosphates	
340	Potassium phosphates	
341	Calcium phosphates	
450	Diphosphates	
451	Triphosphates	

INS no.	Name of additive	Maximum level
452	Polyphosphates	2 000 mg/kg singly or 3 000 mg/kg in combination, expressed as anhydrous substances
500	Sodium carbonates	
501	Potassium carbonates	
Thickener		
407	Carrageenan	150 mg/kg
Emulsifier		
322	Lecithins	Limited by GMP

5 CONTAMINANTS

The products covered by this Standard shall comply with the maximum levels for contaminants that are specified for the product in the *General Standard for Contaminants and Toxins in Food and Feed* (CXS 193-1995).

The milk used in the manufacture of the products covered by this Standard shall comply with the maximum levels for contaminants and toxins specified for milk by the *General Standard for Contaminants and Toxins in Food and Feed* (CXS 193-1995) and with the maximum residue limits for veterinary drug residues and pesticides established for milk by the CAC.

6 HYGIENE

It is recommended that the products covered by the provisions of this standard be prepared and handled in accordance with the appropriate sections of the *General Principles of Food Hygiene* (CXC 1-1969), the *Code of Hygienic Practice for Milk and Milk Products* (CXC 57-2004) and other relevant Codex texts such as Codes of Hygienic Practice and Codes of Practice. The products should comply with any microbiological criteria established in accordance with the *Principles and Guidelines for the Establishment and Application of Microbiological Criteria Related to Foods* (CXG 21-1997).

7 LABELLING

In addition to the provisions of the *General Standard for the Labelling of Prepackaged Foods* (CXS 1-1985) and the *General Standard for the Use of Dairy Terms* (CXS 206-1999), the following specific provisions apply.

7.1 Name of the food

The name of the food shall be:

Sweetened condensed milk	
Sweetened condensed skimmed milk	According to the composition specified in Section 3
Sweetened condensed partly skimmed milk	
Sweetened condensed high-fat milk	

Sweetened condensed partly skimmed milk may be designated "sweetened condensed semi-skimmed milk" if the milkfat content is 4.0%–4.5% and the minimum milk solids is

28% m/m.

7.2 Declaration of milkfat content

If the consumer would be misled by the omission, the milkfat content shall be declared in a manner found acceptable in the country of sale to the final consumer, either (i) as a percentage by mass or volume, or (ii) in grams per serving as quantified in the label provided that the number of servings is stated.

7.3 Declaration of milk protein

If the consumer would be misled by the omission, the milk protein content shall be declared in a manner acceptable in the country of sale to the final consumer, either as (i) a percentage by mass or volume, or (ii) grams per serving as quantified in the label provided the number of servings is stated.

7.4 List of ingredients

Notwithstanding the provision of Section 4.2.1 of the *General Standard for the Labelling of Prepackaged Foods* (CXS 1-1985), milk products used only for protein adjustment need not be declared.

7.5 Labelling of non-retail containers

Information required in Section 7 of this Standards and Sections 4.1 to 4.8 of the *General Standard for the Labelling of Prepackaged Foods* (CXS 1-1985), and, if necessary, storage instructions, shall be given either on the container or in accompanying documents, except that the name of the product, lot identification, and the name and address of the manufacturer or packer shall appear on the container. However, lot identification, and the name and address of the manufacturer or packer may be replaced by an identification mark, provided that such a mark is clearly identifiable with the accompanying documents.

8 METHODS OF SAMPLING and ANALYSIS

For checking the compliance with this standard, the methods of analysis and sampling contained in the *Recommended Methods of Analysis and Sampling* (CXS 234-1999) relevant to the provisions in this standard, shall be used.

干酪通则

GENERAL STANDARD FOR CHEESE

CXS 283-1978

原为CODEX STAN A-6-1973。1973年通过。1999年修订。
2006年、2008年、2010年、2013年、2018年修正。

1　范围

本标准适用于符合本标准第2条所述即食或用于进一步加工的干酪。特殊干酪品种的个体标准或组标准可能含有比本标准更为具体的规定，在符合本标准规定的前提下这些具体规定适用。

2　说明

2.1　干酪：是指成熟或未成熟的半硬质、硬质或特硬质产品，可能有包衣，所含乳清蛋白或酪蛋白比例不超过乳中比例。干酪可通过下面方法制备：

（a）利用凝乳酶或其他适宜的凝结剂的凝结作用，使乳、脱脂乳、部分脱脂乳、奶油、乳清奶油、酪乳或这些乳品任意组合的全部或部分蛋白质凝固，排出凝固作用产生的部分乳清，并遵循以下原则：干酪制作过程会浓缩乳蛋白（尤其是酪蛋白），因而干酪蛋白质含量显著高于制作干酪所采用的上述乳品蛋白质含量；

（b）与乳和/或乳制品相关的蛋白质凝固的加工工艺，其最终产品与（a）中定义的产品具有相似的物理、化学和感官特性。

2.1.1　成熟干酪是一种生产后无法即食的干酪，应在特定温度和条件下储存特定时间，使其产生此类干酪特有的生化和物理变化。

2.1.2　霉菌成熟干酪是一种成熟干酪，主要通过特殊霉菌在干酪内部和/或表面生长来完成成熟过程。

2.1.3　未成熟干酪（包括新鲜干酪）是在生产后短期内即可食用的干酪。

3　基本成分和质量要求

3.1　基本原料

乳和/或来源于乳的各种制品。

3.2　其他配料

— 无害乳酸发酵剂和/或产香菌与其他无害微生物的培养剂；

- 安全、适宜的酶；
- 氯化钠；
- 饮用水。

4 食品添加剂

只允许在规定范围内使用下列各种食品添加剂：

未成熟干酪

按照《未成熟干酪（包括新鲜干酪）标准》（CXS 221-2001）规定。

盐水干酪

按照《盐水干酪标准》（CXS 208-1999）规定。

成熟干酪，包括霉菌成熟干酪

下表中未列出，但在食品法典特殊成熟干酪品种标准中有规定的添加剂，在符合这些标准所规定限值前提下也可用于类似种类的干酪。

INS编号	添加剂名称	最大限量
色素		
100	姜黄素（用于干酪可食用外皮）	根据GMP限量使用
101	核黄素	
120	胭脂红（仅用于红花条纹干酪）	
140	叶绿素（仅用于绿花条纹干酪）	
141	叶绿素铜	15 mg/kg
160a（i）	β-胡萝卜素（人工合成）	25 mg/kg
160a（ii）	β-胡萝卜素（天然萃取物）	600 mg/kg
160b（ii）	胭脂树提取物-红木素	50 mg/kg
160c	辣椒红	根据GMP限量使用
160e	β-apo-8'-胡萝卜素	35 mg/kg
160f	β-apo-8'-胡萝卜素酸乙酯	35 mg/kg
162	甜菜红	根据GMP限量使用
171	二氧化钛	
酸度调节剂		
170	碳酸钙	根据GMP限量使用
504	碳酸镁	
575	葡萄糖酸-δ-内酯	

INS编号	添加剂名称	最大限量
防腐剂		
200	山梨酸	3 000 mg/kg，以山梨酸计
202	山梨酸钾	
203	山梨酸钙	
234	乳酸链球菌素	12.5 mg/kg
239	六亚甲基四胺（仅用于波罗伏洛干酪）	25 mg/kg，以甲醛计
251	硝酸钠	50 mg/kg，以$NaNO_3$计
252	硝酸钾	
280	丙酸	3 000 mg/kg，以丙酸计
281	丙酸钠	
282	丙酸钙	
1105	溶菌酶	根据GMP限量使用
仅适用于表面或外皮处理		
200	山梨酸	1 000 mg/kg，单独或混合使用，以山梨酸计
202	山梨酸钾	
203	山梨酸钙	
235	纳他霉素（匹马菌素）	2 mg/dm^{2}表面，且在5 mm深度不存在
杂类添加剂		
508	氯化钾	根据GMP限量使用
抗结块剂（切块、切片、切碎和磨碎的干酪）		
460	纤维素	根据GMP限量使用
551	二氧化硅（无定型）	10 000 mg/kg，单独或混用，以二氧化硅计算硅酸盐
552	硅酸钙	
553	硅酸镁	
560	硅酸钾	
防腐剂		
200	山梨酸	1 000 mg/kg，单独或混用，以山梨酸计
202	山梨酸钾	
203	山梨酸钙	

5 污染物

本标准所涉及的产品应符合《食品及饲料中污染物和毒素通用标准》（CXS 193-1995）规定的污染物最大限量。

本标准所涉及的产品在加工中使用的牛乳应符合《食品及饲料中污染物和毒素通用标准》（CXS 193-1995）规定的乳中污染物和毒素最大限量以及国际食品法典委员会设定的农药和兽药最大残留限量。

6 卫生要求

建议本标准所涉及的产品应遵循《食品卫生总则》（CXC 1-1969）、《乳及乳制品卫生操作规范》（CXC 57-2004）以及《卫生操作规范》和《生产操作规范》等其他相关法典文本。产品应符合《食品微生物标准制定与实施原则和准则》（CXG 21-1997）规定的所有微生物标准。

7 标识

除符合《预包装食品标识通用标准》（CXS 1-1985）和《乳品术语》（CXS 206-1999）外，还应符合下列具体标准。

7.1 产品名称

产品名称应为干酪。但一些特殊干酪品种的名称中可以省略"干酪"一词，前提是国际食品法典中有该干酪品种的标准或在无此标准的情况下产品销售国法规中已对该产品指定具体名称，且此种省略不会使消费者对产品特性产生误解。

7.1.1 若产品无品种名称而仅命名为"干酪"，则可附加下表给出适当描述性用语。

根据硬度和成熟特性命名		
根据硬度：术语1		根据主要的成熟方法：术语2
含量（%）	命名	
<51	特硬	成熟
49~56	硬质	霉菌成熟
54~69	坚实/半硬	未成熟/新鲜
>67	软质	盐水浸泡

MFFB指水分占干酪无脂总重百分比，即

$$\frac{干酪中水分重量}{干酪总重-干酪中脂肪重量} \times 100$$

例如：

某干酪中水分占其无脂总重57%，其成熟方法与丹麦蓝纹干酪（Danablu）相似，则该干酪可命名为："霉菌成熟半硬质干酪或半硬质霉菌成熟干酪"。

7.2 乳脂含量说明

乳脂含量应以销售国可接受的方式向最终消费者声明：（i）质量百分比形式；（ii）干物质中脂肪百分比形式；（iii）如果产品标识上标明了份数，也可用每份中乳脂重量（g）表示。

另外，可使用下列术语：

高脂	（如果干物质脂肪含量≥60%）
全脂	（如果干物质脂肪含量≥45%，<60%）
中脂	（如果干物质脂肪含量≥25%，<45%）
部分脱脂	（如果干物质脂肪含量≥10%，<25%）
脱脂	（如果干物质脂肪含量<10%）

7.3 日期标识

虽然《预包装食品标识通用标准》（CXS 1-1985）第4.7.1条有规定，但半硬、硬质和特硬干酪不必在标签中注明保质期，此类干酪并非霉菌成熟干酪或软质成熟干酪，也不是售卖给最终消费者的：这种情况下只需标明生产日期。

7.4 非零售包装标识

在包装容器上除要标明产品名称、批次、生产厂家或包装商的名称和地址外，在包装容器上或附随说明书中，也应按本标准第7条和《预包装食品标识通用标准》（CXS 1-1985）第4.1～4.8条所要求的信息以及储存说明（在必要的地方）加以陈述。如果批次、生产厂家或包装商的名称和地址可以在附随的文件标明，则可以用一个识别标识来代替。

8 抽样和分析方法

为检查产品是否符合本标准，应采用《分析和抽样推荐性方法》（CXS 234-1999）中涉及本标准规定的分析和抽样方法。

附录

附 加 信 息[1]

干酪外皮

在自然环境下或在空气湿度受控、空气成分（可能）受控环境下，含霉菌干酪凝块在成熟过程中其外围会形成水分含量较低的半封闭层，即所谓外皮。构成外皮的干酪部分在成熟开始时与干酪内部有相同成分。很多情况下干酪外皮形成始于浓盐水浸泡。外皮受到盐水中盐梯度、氧气、干燥和其他反应影响，其成分逐渐变得与干酪内部不同，味道通常更苦。

在干酪成熟期间或之后，可对外皮处理或引入良种微生物培养物，例如青霉菌或亚麻短杆菌，通过自然繁殖处理外皮。处理后形成层有时被称为黏液，也是外皮一部分。无外皮干酪的成熟使用了成熟膜。干酪外部不会形成水分含量较低外皮，当然光照影响还是会使干酪外部与内部略有不同。

干酪表面

"干酪表面"一词用于描述干酪或碎干酪（即使是切片、切碎或磨碎的干酪）外层。该术语包含整个干酪外部，无论干酪外皮是否已经形成。

干酪包衣

干酪可在成熟前、成熟过程中或成熟完成后包衣。在成熟过程中包衣的目的是调节干酪水分含量，保护干酪不受微生物侵害。

在成熟完成后包衣目的是保护干酪免遭微生物侵害和其他污染，防止干酪在运输和经销过程中受到物理损害，使干酪呈现特定外观（如颜色）。

包衣与外皮很容易区分，因为包衣由非干酪材料制成且通常可以刷掉、揉掉或剥掉。干酪可以用下列材料包衣：

- 通常以聚乙烯乙酸酯为材料制成的膜，也可是其他人造材料或天然配料制成的材料，在成熟过程中膜有助于调节湿度，保护干酪免遭微生物侵害（如熟化膜）[2]。
- 主要由蜡、石蜡或塑料制成的覆盖层，通常不透水，在成熟后保护干酪免遭微生物侵害，在零售流通中防止干酪受到物理损害，有时还用于保持干酪良好外观。

---

1 该修正由食品法典委员会第 26 届会议通过（2003）。
2 出于技术原因，麸麸或小麦蛋白产品不应用作天然不含麸质食品的包衣或加工助剂——《小麦蛋白产品（包括麦麸）标准》（CXS 163-1987）。

GENERAL STANDARD FOR CHEESE

CXS 283-1978

Formerly CODEX STAN A-6-1973. Adopted in 1973. Revised in 1999.

Amended in 2006, 2008, 2010, 2013, 2018.

1 SCOPE

This Standard applies to all products, intended for direct consumption or further processing, in conformity with the definition of cheese in Section 2 of this Standard. Subject to the provisions of this Standard, standards for individual varieties of cheese, or groups of varieties of cheese, may contain provisions which are more specific than those in this Standard and in these cases, those specific provisions shall apply.

2 DESCRIPTION

2.1 Cheese is the ripened or unripened soft, semi-hard, hard, or extra-hard product, which may be coated, and in which the whey protein/casein ratio does not exceed that of milk, obtained by:

(a) coagulating wholly or partly the protein of milk, skimmed milk, partly skimmed milk, cream, whey cream or buttermilk, or any combination of these materials, through the action of rennet or other suitable coagulating agents, and by partially draining the whey resulting from the coagulation, while respecting the principle that cheese-making results in a concentration of milk protein (in particular, the casein portion), and that consequently, the protein content of the cheese will be distinctly higher than the protein level of the blend of the above milk materials from which the cheese was made; and/or

(b) processing techniques involving coagulation of the protein of milk and/or products obtained from milk which give an end-product with similar physical, chemical and organoleptic characteristics as the product defined under (a).

2.1.1 Ripened cheese is cheese which is not ready for consumption shortly after manufacture but which must be held for such time, at such temperature, and under such other conditions as will result in the necessary biochemical and physical changes characterizing the cheese in question.

2.1.2 Mould ripened cheese is a ripened cheese in which the ripening has been accomplished primarily by the development of characteristic mould growth throughout the interior and/or on the surface of the cheese.

2.1.3 Unripened cheese including fresh cheese is cheese which is ready for consumption shortly after manufacture.

3 ESSENTIAL COMPOSITION AND QUALITY FACTORS

3.1 Raw materials

Milk and/or products obtained from milk.

3.2 Permitted ingredients

– Starter cultures of harmless lactic acid and/or flavour producing bacteria and cultures of other harmless microorganisms;
– Safe and suitable enzymes;
– Sodium chloride;
– Potable water.

4 FOOD ADDITIVES

Only those food additives listed below may be used and only within the limits specified.

Unripened cheeses

As listed in the *Group Standard for Unripened Cheese Including Fresh Cheese* (CXS 221-2001).

Cheeses in brine

As listed in the *Standard for Cheeses in Brine* (CXS 208-1999).

Ripened cheeses, including mould ripened cheeses

Additives not listed below but provided for in Codex individual standards for varieties of ripened cheeses may also be used for similar types of cheese within the limits specified within those standards.

INS no.	Name of additive	Maximum level
Colours		
100	Curcumins (for edible cheese rind)	Limited by GMP
101	Riboflavins	
120	Carmines (for red marbled cheeses only)	
140	Chlorophyll (for green marbled cheeses only)	
141	Chlorophylls, copper complexes	15 mg/kg
160a(i)	Carotene, *beta*-, synthetic	25 mg/kg
160a(ii)	Carotenes, *beta*-, vegetable	600 mg/kg
160b(ii)	Annatto extracts, norbixin based	50 mg/kg
160c	Paprika oleoresin	Limited by GMP
160e	Carotenal, *beta-apo*-8'-	35 mg/kg
160f	Carotenoic acid, ethyl ester, *beta- apo*-8'-	35 mg/kg
162	Beet red	Limited by GMP
171	Titanium dioxide	
Acidity regulators		
170	Calcium carbonates	Limited by GMP
504	Magnesium carbonates	
575	Glucono *delta*-lactone	

INS no.	Name of additive	Maximum level
Preservatives		
200	Sorbic acid	3 000 mg/kg calculated as sorbic acid
202	Potassium sorbate	
203	Calcium sorbate	
234	Nisin	12.5 mg/kg
239	Hexamethylene tetramine (Provolone only)	25 mg/kg, expressed as formaldehyde
251	Sodium nitrate	50 mg/kg, expressed as NaNO3
252	Potassium nitrate	
280	Propionic acid	3 000 mg/kg, calculated as propionic acid
281	Sodium propionate	
282	Calcium propionate	
1105	Lysozyme	Limited by GMP
For surface/rind treatment only:		
200	Sorbic acid	1 000 mg/kg singly or in combination, calculated as sorbic acid
202	Potassium sorbate	
203	Calcium sorbate	
235	Natamycin (pimaricin)	2 mg/dm^2 of surface. Not present in a depth of 5 mm
Miscellaneous additive		
508	Potassium chloride	Limited by GMP
Anti-caking agents (Sliced, cut, shredded or grated cheese)		
460	Celluloses	Limited by GMP
551	Silicon dioxide, amorphous	10 000 mg/kg singly or in combination. Silicates calculated as silicon dioxide
552	Calcium silicate	
553	Magnesium silicates	
560	Potassium silicate	
Preservatives		
200	Sorbic acid	1 000 mg/kg singly or in combination, calculated as sorbic acid
202	Potassium sorbate	
203	Calcium sorbate	

5 CONTAMINANTS

The products covered by this Standard shall comply with the Maximum Levels for contaminants that are specified for the product in the *General Standard for Contaminants and Toxins in Food and Feed* (CXS 193-1995).

The milk used in the manufacture of the products covered by this Standard shall comply with the Maximum Levels for contaminants and toxins specified for milk by the *General Standard for Contaminants and Toxins in Food and Feed* (CXS 193-1995) and with the maximum residue limits for veterinary drug residues and pesticides established for milk by the CAC.

6 HYGIENE

It is recommended that the products covered by the provisions of this standard be prepared and handled in accordance with the appropriate sections of the *General Principles of Food Hygiene* (CXC 1-1969), the *Code of Hygienic Practice for Milk and Milk Products* (CXC 57-2004) and other relevant Codex texts such as Codes of Hygienic Practice and Codes of Practice. The products should comply with any microbiological criteria established in accordance with the *Principles and Guidelines for the Establishment and Application of Microbiological Criteria Related to Foods* (CXG 21-1997).

7 LABELLING

In addition to the provisions of the General *Standard for the Labelling of Prepackaged Foods* (CXS 1-1985) and the *General Standard for the Use of Dairy Terms* (CXS 206-1999), the following specific provisions apply.

7.1 Name of the food

The name of the food shall be cheese. However, the word "cheese" may be omitted in the designation of an individual cheese variety reserved by a Codex standard for individual cheeses, and, in the absence thereof, a variety name specified in the national legislation of the country in which the product is sold, provided that the omission does not create an erroneous impression regarding the character of the food.

7.1.1

In case the product is not designated with a variety name but with the designation "cheese" alone, the designation may be accompanied by the appropriate descriptive terms in the following table:

DESIGNATION ACCORDING TO FIRMNESS AND RIPENING CHARACTERISTICS		
According to firmness: Term 1		According to principal ripening: Term 2
MFFB (%)	Designation	
< 51	Extra hard	Ripened
49–56	Hard	Mould ripened
54–69	Firm/Semi-hard	Unripened/Fresh
>67	Soft	In Brine

MFFB equals percentage moisture on a fat-free basis, i.e.,

$$\frac{\text{Weight of moisture in the cheese}}{\text{Total weight of cheese} - \text{Weight of fat in the cheese}} \times 100$$

Example:

The designation of a cheese with moisture on a fat-free basis of 57% which is ripened in a manner similar in which Danablu is ripened would be:

"Mould ripened firm cheese or firm mould ripened cheese."

7.2 Declaration of milk fat content

The milk fat content shall be declared in a manner found acceptable in the country of sale to the final consumer, either (i) as a percentage by mass, (ii) as a percentage of fat in dry matter, or (iii) in grams per serving as quantified in the label provided that the number of servings is stated.

Additionally, the following terms may be used:

High fat	(if the content of FDM is above or equal to 60%)
Full fat	(if the content of FDM is above or equal to 45% and less than 60%)
Medium fat	(if the content of FDM is above or equal to 25% and less than 45%)
Partially skimmed	(if the content of FDM is above or equal to 10% and less than 25%)
Skim	(if the content of FDM is less than 10%)

7.3 Date marking

Notwithstanding the provisions of Section 4.7.1 of the *General Standard for the Labelling of Prepackaged Foods* (CXS 1-1985), the date of minimum durability need not be declared in the labelling of firm, hard and extra hard cheese which are not mould/soft-ripened and not intended to be purchased as such by the final consumer: in such cases the date of manufacture shall be declared.

7.4 Labelling of non-retail containers

Information required in Section 7 of this Standard and Sections 4.1 to 4.8 of the *General Standard for the Labelling of Prepackaged Foods* (CXS 1-1985), and, if necessary, storage instructions, shall be given either on the container or in accompanying documents, except that the name of the product, lot identification, and the name and address of the manufacturer or packer shall appear on the container, and in the absence of such a container on the cheese itself. However, lot identification, and the name and address of the manufacturer or packer may be replaced by an identification mark, provided that such a mark is clearly identifiable with the accompanying documents.

8 METHODS OF SAMPLING AND ANALYSIS

For checking the compliance with this standard, the methods of analysis and sampling contained in the *Recommended Methods of Analysis and Sampling* (CXS 234-1999) relevant to the provisions in this standard, shall be used.

APPENDIX[1]

CHEESE RIND

During ripening of the moulded cheese curd in natural creation or in environments in which the air humidity and, possibly, air composition are controlled, the outside of the cheese will develop into a semi-closed layer with a lower moisture content. This part of the cheese is called **rind**. The rind is constituted of cheese mass which, at the start of the ripening, is of the same composition as the internal part of the cheese. In many cases, the brining of cheese initiates the formation of rind. Due to the influence of the salt gradient in the brine, of oxygen, of drying out and of other reactions, the rind successively becomes of a somewhat different composition than the interior of the cheese and often presents a more bitter taste.

During or after ripening the cheese rind can be treated or can be naturally colonized with desired cultures of microorganisms, for instance *Penicillium candidum* or *Brevibacterium linens*. The resulting layer, in some cases referred to as **smear**, forms a part of the rind.

Rindless cheese is ripened by the use of a ripening film. The outer part of that cheese does not develop a rind with a lower moisture content although influence of light of course can cause some difference compared to the inner part.

CHEESE SURFACE

The term "**cheese surface**" is used for the outside layer of cheese or parts of cheese, even in the sliced, shredded or grated form. The term includes the outside of the whole cheese, disregarding whether a rind has been formed or not.

CHEESE COATINGS

Cheese can be coated prior to the ripening, during the ripening process or when the ripening has been finished. When a coating is used during ripening the purpose of the coating is to regulate the moisture content of the cheese and to protect the cheese against micro-organisms.

Coating of a cheese after the ripening has been finished is done to protect the cheese against microorganisms and other contamination, to protect the cheese from physical damage during transport and distribution and/or to give the cheese a specific appearance (e.g. coloured).

Coating can be distinguished very easily from rind, as coatings are made of non-cheese material, and very often it is possible to remove the coating again by brushing, rubbing or peeling it off.

Cheese can be coated with:

- A film, very often polyvinylacetate, but also other artificial material or material composed of natural ingredients, which helps to regulate the humidity during ripening and protects the cheese against microorganisms (for example, ripening

---

1 Amendment adopted by the 26th Session of the Codex Alimentarius Commission (2003).

films).[1]

- layer, mostly wax, paraffin or a plastic, which normally is impermeable to moisture, to protect the cheese after ripening against microorganisms and against physical damage during retail handling and, in some cases to contribute to the presentation of the cheese.

---

[1] Wheat gluten or wheat protein products should not be used for technological reasons e.g. coating or processing aids for foods which are gluten-free by nature – *Standard for Wheat Protein Products including Wheat Gluten* (CXS 163-1987).

乳 清 干 酪

STANDARD FOR WHEY CHEESE

CXS 284-1971

前为CODEX STAN A-7-1971。1971年通过。

1999年、2006年修订。2010年、2018年修正。

1　范围

本标准适用于符合本标准第2条所述即食或用于进一步加工的乳清干酪。就本标准规定而言,乳清干酪个别品种食典标准规定可能细于本标准规定。

2　说明

2.1　乳清干酪是主要通过以下方式加工而成的固态、半固态或软质制品:

（1）乳清浓缩和浓缩制品成型;

（2）乳清热凝固,加或不加酸均可。

无论采用哪种方式,乳清都可先行预浓缩,后进行乳清浓缩或乳清蛋白凝固。在该过程中,也可在浓缩或凝固前后添加乳、稀奶油或其他乳原料。乳清凝固制品中乳清蛋白与酪蛋白比例应显著高于乳中比例。

乳清凝固可以制成成熟或非成熟制品。

2.2　乳清浓缩而成的乳清干酪通过乳清或乳清与乳、稀奶油或其他乳原料混合物热蒸发制成,浓度应使干酪成品保持稳定形状。由于乳糖含量较高,此类干酪色泽一般在淡黄色与褐色之间,具有甜味、熟味或焦糖味。

2.3　乳清凝固而成的乳清干酪通过乳清,或者乳清与乳或稀奶油的混合物热沉淀制成,加不加酸均可。此类乳清干酪乳糖含量较低,色泽在白色与淡黄色之间。

3　基本成分和质量要求

3.1　基本原料

（1）用于生产乳清浓缩制品:乳清、稀奶油、乳和其他乳原料。

（2）用于生产乳清凝固制品:乳清、乳、稀奶油和酪乳。

3.2　其他配料

仅用于乳清凝固制品:

— 氯化钠;

— 无害乳酸菌发酵剂。

仅用于乳清热处理浓缩制品：
- 糖类（以GMP为限）。

3.3 允许营养素

按照《食品中必需营养素添加通用原则》（CXG 9-1987）相关规定，应根据各个国家需求，酌情通过国家立法制定矿物质和其他营养素的最高和最低限量，包括酌情禁用某些营养素。

4 食品添加剂

可以根据本标准在食品中使用《食品添加剂通用标准》（CXS 192-1995）表1和表2食品类别01.6.3（乳清干酪）和01.6.6（乳清蛋白干酪）所列食品添加剂。

5 污染物

本标准所涉及的产品应符合《食品及饲料中污染物和毒素通用标准》（CXS 193-1995）规定的污染物最大限量。

本标准所涉及的产品在加工中使用的牛乳应符合《食品及饲料中污染物和毒素通用标准》（CXS 193-1995）规定的乳中污染物和毒素最大限量以及国际食品法典委员会设定的农药和兽药最大残留限量。

6 卫生要求

建议本标准所涉及的产品应遵循《食品卫生总则》（CXC 1-1969）、《乳及乳制品卫生操作规范》（CXC 57-2004）以及《卫生操作规范》和《生产操作规范》等其他相关法典文本。产品应符合《食品微生物标准制定与实施原则和准则》（CXG 21-1997）规定的所有微生物标准。

7 标识

除符合《预包装食品标识通用标准》（CXS 1-1985）和《乳品术语》（CXS 206-1999）外，还应符合以下具体规定。

7.1 产品名称

产品名称应为乳清干酪。如产品销售国认为消费者必须知情，则必须说明产品性质。对于个别干酪食典标准备用的乳清干酪个别品种的名称和产品销售国法律规定的品种名称（如无相关食典标准），可以略去"乳清干酪"，但不应使消费者对产品特性产生误解。

如乳清凝固而成的乳清干酪不以产品名称命名，而以"乳清干酪"命名，则可在名称中加入《干酪通则》（CXS 283-1978）第7.1.1条规定的形容词。

乳清浓缩而成的非成熟乳清干酪可以按照第7.2条规定，根据脂肪含量命名。

7.2　乳脂含量说明

乳脂含量应以销售国可接受的方式向最终消费者声明：（i）质量百分比形式；（ii）干物质中脂肪百分比形式；（iii）如果产品标识上标明了份数，也可用每份中乳脂重量（g）表示。

对乳清浓缩而成的干酪，乳脂含量声明可以加注脂肪含量，限量如下：

干物质脂肪含量[1]	
加稀奶油的乳清干酪	≥33%
乳清干酪	≥10%，<33%
脱脂乳清干酪	<10%

7.3　非零售包装标识

在包装容器上除要标明产品名称、批次、生产厂家或包装商的名称和地址外，在包装容器上或附随说明书中，也应按本标准第7条和《预包装食品标识通用标准》（CXS 1-1985）第4.1～4.8条所要求的信息以及储存说明（在必要的地方）加以陈述。如果批次、生产厂家或包装商的名称和地址可以在附随的文件标明，则可以用一个识别标识来代替。

8　抽样和分析方法

为检查产品是否符合本标准，应采用《分析和抽样推荐性方法》（CXS 234-1999）中涉及本标准规定的分析和抽样方法。

1　乳清干酪干物质含量包括乳糖结晶水。

STANDARD FOR WHEY CHEESES

CXS 284-1971

Formerly CODEX STAN A-7-1971. Adopted in 1971.

Revised in 1999, 2006. Amended in 2010, 2018.

1 SCOPE

This Standard applies to all products intended for direct consumption or further processing, in conformity with the definition of whey cheeses in Section 2 of this Standard. Subject to the provisions of this Standard, Codex standards for individual varieties of whey cheeses may contain provisions which are more specific than those in this Standard.

2 DESCRIPTION

2.1

Whey Cheeses are solid, semi-solid, or soft products which are principally obtained through either of the following processes:

(1) he concentration of whey and the moulding of the concentrated product;

(2) the coagulation of whey by heat with or without the addition of acid.

In each case, the whey may be pre-concentrated prior to the further concentration of whey or coagulation of the whey proteins. The process may also include the addition of milk, cream, or other raw materials of milk origin before or after concentration or coagulation. The ratio of whey protein to casein in the product obtained through the coagulation of whey shall be distinctly higher than that of milk.

The product obtained through the coagulation of whey may either be ripened or unripened.

2.2

Whey Cheese obtained through the concentration of whey is produced by heat evaporation of whey, or a mixture of whey and milk, cream, or other raw materials of milk origin, to a concentration enabling the final cheese to obtain a stable shape. Due to their relatively high lactose content these cheeses are typically yellowish to brown in colour and possess a sweet, cooked, or caramelized flavour.

2.3

Whey Cheese obtained through the coagulation of whey is produced by heat precipitation of whey, or a mixture of whey and milk or cream, with or without the addition of acid. These whey cheeses have a relatively low lactose content and a white to yellowish colour.

3 ESSENTIAL COMPOSITION AND QUALITY FACTORS

3.1 Raw materials

(1) For products obtained through the concentration of whey:

whey, cream, milk and other raw materials obtained from milk.

(2) For products obtained through the coagulation of whey:

whey, milk, cream and buttermilk.

3.2 **Permitted ingredients**

Only for use in products obtained by coagulation of whey:

- Sodium chloride;
- Starter cultures of harmless lactic acid bacteria.

Only for use in products obtained through the concentration of whey by heat treatment

- Sugars (limited by GMP).

3.3 **Permitted nutrients**

Where allowed in accordance with the *General Principles for the Addition of Essential Nutrients to Foods* (CXG 9-1987), maximum and minimum levels for minerals and other nutrients, where appropriate, should be laid down by national legislation in accordance with the needs of individual country including, where appropriate, the prohibition of the use of particular nutrients.

4 **FOOD ADDITIVES**

Food additives listed in Tables 1 and 2 of the *General Standard for Food Additives* (CXS 192-1995) in Food Category 01.6.3 (Whey cheese) and 01.6.6 (Whey protein cheese) may be used in foods subject to this Standard.

5 **CONTAMINANTS**

The products covered by this Standard shall comply with the maximum levels for contaminants that are specified for the product in the *General Standard for Contaminants and Toxins in Food and Feed* (CXS 193-1995).

The milk used in the manufacture of the products covered by this Standard shall comply with the maximum levels for contaminants and toxins specified for milk by the *General Standard for Contaminants and Toxins in Food and Feed* (CXS 193-1995) and with the maximum residue limits for veterinary drug residues and pesticides established for milk by the CAC.

6 **HYGIENE**

It is recommended that the products covered by the provisions of this standard be prepared and handled in accordance with the appropriate sections of the *General Principles of Food Hygiene* (CXC 1-1969), the *Code of Hygienic Practice for Milk and Milk Products* (CXC 57-2004) and other relevant Codex texts such as Codes of Hygienic Practice and Codes of Practice. The products should comply with any microbiological criteria established in accordance with the *Principles and Guidelines for the Establishment and Application of Microbiological Criteria Related to Foods* (CXG 21-1997).

7 **LABELLING**

In addition to the provisions of the *General Standard for the Labelling of Prepackaged Foods* (CXS 1-1985) and the *General Standard for the Use of Dairy Terms* (CXS 206-1999), the following specific provisions apply.

7.1 **Name of the food**

The name of the food shall be **whey cheese**. Where it is considered necessary for consumer information in the country of sale, a description of the nature of the product may be required. The words "whey cheese" may be omitted in the designation of an individual whey cheese variety reserved by a Codex standard for individual cheeses, and, in the absence thereof, a variety name specified in the national legislation of the country in which the product is sold, provided that the omission does not create an erroneous impression regarding the character of the food.

In case a whey cheese obtained through the co-agulation of whey is not designated by a variety name, but with the designation "whey cheese", the designation may be accompanied by a descriptive term such as provided for in Section 7.1.1 of the *General Standard for Cheese* (CXS 283-1978).

Unripened whey cheese obtained through the concentration of whey may be designated according to the fat content as provided in Section 7.2.

7.2 Declaration of milk fat content

The milk fat content shall be declared in a manner found acceptable in the country of sale to the final consumer, either (i) as a percentage by mass, (ii) as a percentage of fat in dry matter, or (iii) in grams per serving as quantified in the label provided that the number of servings is stated.

For cheeses obtained from the concentration of whey, the declaration of milk fat content may be combined with an indication of the fat content as follows:

Fat on the dry basis[1]

Creamed whey cheese	minimum 33%
Whey cheese	minimum 10% and less than 33%
Skimmed whey cheese	less than 10%

7.3 Labelling of non-retail containers

Information required in Section 7 of this Standard and Sections 4.1 to 4.8 of the *General Standard for the Labelling of Prepackaged Foods* (CXS 1-1985), and, if necessary, storage instructions, shall be given either on the container or in accompanying documents, except that the name of the product, lot identification, and the name and address of the manufacturer or packer shall appear on the container. However, lot identification, and the name and address of the manufacturer or packer may be replaced by an identification mark, provided that such a mark is clearly identifiable with the accompanying documents.

8 METHODS OF SAMPLING AND ANALYSIS

For checking the compliance with this Standard, the methods of analysis and sampling contained in the *Recommended Methods of Analysis and Sampling* (CXS 234-1999) relevant to the provisions in this Standard, shall be used.

---

[1] The dry matter content of whey cheese includes water of crystallization of the lactose.

稀奶油和预制稀奶油

STANDARD FOR CREAM AND PREPARED CREAMS

CXS 288-1976

前为CODEX STAN A-9-1976。1976年通过。

2003年、2008年修订。2010年、2018年修正。

1 范围

本标准适用于符合本标准第2条所述即食或用于进一步加工的稀奶油和预制稀奶油。

2 说明

2.1 稀奶油是指液态[1]乳制品,脂肪含量较高,呈乳剂状,从乳中将脱脂乳脂肪物理分离而成。

2.2 复原稀奶油是指乳制品复原而成的稀奶油,加或不加饮用水均可,最终产品特性与第2.1条所述产品特性一致。

2.3 调配稀奶油是指乳制品调配而成的稀奶油,加或不加饮用水均可,最终产品特性与第2.1条所述产品特性一致。

2.4 预制稀奶油是指对稀奶油、复原稀奶油和/或调配稀奶油适当加工处理而成的乳制品,旨在获得下述特性。

2.4.1 预包装液态奶油是指稀奶油、复原稀奶油和/或调配稀奶油预制和包装而成的液态[1]乳制品,供直接食用和/或直接使用。

2.4.2 搅打稀奶油是指用于搅打的液态[1]稀奶油、复原稀奶油和/或调配稀奶油。如供最终消费者使用,则稀奶油应经预制,方便搅打加工。

2.4.3 加压包装稀奶油是指液态[1]稀奶油、复原稀奶油和/或调配稀奶油,与气体推进剂装入推压容器,一出容器即成搅打的稀奶油。

2.4.4 搅打稀奶油是指液态[1]稀奶油、复原稀奶油和/或调配稀奶油,注入空气或惰性气体,但不改变脱脂乳脂肪的乳剂形态。

2.4.5 发酵稀奶油是指稀奶油、复原稀奶油或调配稀奶油经适当微生物作用发酵而成的乳制品,最终pH值降低,呈凝固或非凝固状。如标识直接、或间接标注具体微生物的含量或在销售环节以含量声明的形式标注,则在保质期内相关微生物应在产品中存活并具活性,数量丰富。如产品在发酵后经热处理,则不适用活性微生物的要求。

2.4.6 酸化稀奶油是指稀奶油、复原稀奶油和/或调配稀奶油经酸和/或酸度调节剂作用酸

---

[1] 液态是指可在冻结温度以上倾倒。

化而成的乳制品，最终pH值降低，呈凝固或非凝固状。

3 基本成分和质量要求

3.1 基本原料

用于生产的所有稀奶油和预制稀奶油：

乳，可在加工稀奶油前进行机械和物理处理。

此外，用于生产的复原稀奶油或调配稀奶油：

黄油[1]、乳脂制品[1]、乳粉[1]、稀奶油粉[1]和饮用水。

此外，用于生产符合第2.4.2～2.4.6条说明的预制稀奶油：

黄油和乳脂制品（通常称为酪乳）生产过程中，乳和稀奶油搅拌分离乳脂后留存并可能已经浓缩和/或干燥的产品。

3.2 其他配料

只有下表所列配料可用于指定用途和产品类别，并以指定限量为限。

如用于仅可合理使用稳定剂和/或增稠剂的产品（见第4条列表）：

- 完全从乳或乳清中提取且各类乳蛋白（包括酪蛋白和乳清蛋白制品和浓缩物及其任意搭配的混合物）含量≥35%（m/m）的产品和乳粉：这类产品可以替代增稠剂和稳定剂发挥相同功能，但须斟酌第4条所列稳定剂和增稠剂的一切用法，添加剂量仅以发挥必要功能为限，不超过20 g/kg；
- 明胶和淀粉：这类物质可以替代稳定剂发挥相同功能，但应考虑第4条所列稳定剂/增稠剂的一切用法，添加剂量以良好生产规范为准，仅以发挥必要的功能为限。

此外，如用于发酵稀奶油，仅可使用：

- 无害微生物发酵剂，包括《发酵乳制品》（CXS 243-2003）第2条规定的发酵剂。

此外，如用于发酵稀奶油和酸化稀奶油，仅可使用：

- 凝乳酶和其他安全、适宜的凝固酶，在不形成酶凝固的前提下提升口感；
- 氯化钠。

3.3 成分

乳脂：最低10%（w/w）。

成分调整低于上述乳脂最小限量，视为不符合《乳品术语》（CXS 206-1999）第4.3.3条规定。

4 食品添加剂

只有下表所列添加剂类别才可用于指定产品类别。在下表允许使用的各类添加剂

---

[1] 详见相关食典标准。

中，只可使用下表所列添加剂，并以指定限量为限。

包括改性淀粉在内，可以根据乳制品定义，单独或混合使用稳定剂和增稠剂，但须斟酌第3.2条规定的明胶和淀粉的一切用法，仅以发挥必要的功能为限。

产品类别	添加剂功能类别			
	稳定剂[a]	酸度调节剂[a]	增稠剂[a]和乳化剂[a]	包装气体和推进剂
预包装液态奶油（2.4.1）	×	×	×	－
搅打稀奶油（2.4.2）	×	×	×	－
加压包装稀奶油（2.4.3）	×	×	×	×
搅打的稀奶油（2.4.4）	×	×	×	×
发酵稀奶油（2.4.5）	×	×	×	－
酸化稀奶油（2.4.6）	×	×	×	－

（a）如需保证产品稳定性和乳剂完整性，则可以参考脂肪含量和产品有效期，使用这类添加剂。关于有效期，鉴于某些低度巴氏杀菌产品无须使用某些添加剂，应特别考虑热处理程度。

× 使用该类别添加剂符合技术要求。

－ 使用该类别添加剂不符合技术要求。

INS编号	添加剂名称	最大限量
酸度调节剂		
270	乳酸（L-，D-和DL-）	根据GMP限量使用
325	乳酸钠	
326	乳酸钾	
327	乳酸钙	
330	柠檬酸	
333	柠檬酸钙	
500（i）	碳酸钠	
500（ii）	碳酸氢钠	
500（iii）	倍半碳酸钠	
501（i）	碳酸钾	
501（ii）	碳酸氢钾	

INS编号	添加剂名称	最大限量
稳定剂和增稠剂		
170（i）	碳酸钙	根据GMP限量使用
331（i）	柠檬酸二氢钠	
331（iii）	柠檬酸钠	
332（i）	柠檬酸二氢钾	
332（ii）	柠檬酸钾	
516	硫酸钙	
339（i）	磷酸二氢钠	1 100 mg/kg，以磷计
339（ii）	磷酸氢二钠	
339（iii）	磷酸三钠	
340（i）	磷酸二氢钾	
340（ii）	磷酸氢二钾	
340（iii）	磷酸三钾	
341（i）	磷酸二氢钙	
341（ii）	磷酸氢二钙	
341（iii）	磷酸三钙	
450（i）	磷酸二钠	
450（ii）	焦磷酸三钠	
450（iii）	焦磷酸四钠	
450（v）	二磷酸四钾	
450（vi）	磷酸二钙	
450（vii）	二磷酸二氢钙	
451（i）	三磷酸五钠	
451（ii）	三磷酸五钾	
452（i）	六偏磷酸钠	
452（ii）	聚磷酸钾	
452（iii）	聚磷酸钠钙	
452（iv）	聚磷酸钙	
452（v）	聚磷酸铵	

INS编号	添加剂名称	最大限量
400	海藻酸	根据GMP限量使用
401	海藻酸钠	
402	海藻酸钾	
403	海藻酸铵	
404	海藻酸钙	
405	海藻酸丙二醇酯	5 000 mg/kg
406	琼脂	根据GMP限量使用
407	卡拉胶	
407a	加工琼芝属海藻胶（PES）	
410	槐豆胶（又名刺槐豆胶）	
412	瓜尔胶	
414	阿拉伯胶	
415	黄原胶	
418	结冷胶	
440	果胶	
460（i）	微晶纤维素（纤维素凝胶）	
460（ii）	纤维素粉	
461	甲基纤维素	
463	羟丙基纤维素	
464	羟丙基甲基纤维素	
465	甲基乙基纤维素	
466	羧甲基纤维素钠（纤维素胶）	
472e	双乙酰酒石酸单、双甘油酯	5 000 mg/kg
508	氯化钾	根据GMP限量使用
509	氯化钙	
1410	单淀粉磷酸酯	
1412	磷酸酯双淀粉	
1413	磷酸化二淀粉磷酸酯	
1414	乙酰化二淀粉磷酸酯	

INS编号	添加剂名称	最大限量
1420	醋酸酯淀粉	根据GMP限量使用
1422	乙酰化双淀粉己二酸酯	
1440	羟丙基淀粉	
1442	羟丙基二淀粉磷酸酯	
1450	辛烯基琥珀酸淀粉钠	
乳化剂		
322（i）	卵磷脂	根据GMP限量使用
432	聚氧乙烯（20）山梨糖醇酐单月桂酸酯	1 000 mg/kg
433	聚氧乙烯（20）山梨醇酐单油酸酯	
434	聚氧乙烯（20）山梨醇酐单棕榈酸酯	
435	聚氧乙烯（20）山梨醇酐单硬脂酸酯	
436	聚氧乙烯（20）山梨糖醇酐三硬脂酸酯	
471	单、双甘油脂肪酸酯	根据GMP限量使用
472a	乙酰化单、双甘油脂肪酸酯	
472b	乳酸脂肪酸甘油酯	
472c	柠檬酸脂肪酸甘油酯	
473	蔗糖脂肪酸酯	5 000 mg/kg
475	聚甘油脂肪酸酯	6 000 mg/kg
491	山梨醇酐单硬脂酸酯	5 000 mg/kg
492	山梨醇酐三硬脂酸酯	
493	山梨醇酐单月桂酸酯	
494	山梨醇酐单油酸酯	
495	山梨醇酐单棕榈酸酯	
包装气体		
290	二氧化碳	根据GMP限量使用
941	氮气	
推进剂		
942	一氧化二氮	根据GMP限量使用

5 污染物

本标准所涉及的产品应符合《食品及饲料中污染物和毒素通用标准》（CXS 193-1995）规定的污染物最大限量。

本标准所涉及的产品在加工中使用的牛乳应符合《食品及饲料中污染物和毒素通用标准》（CXS 193-1995）规定的乳中污染物和毒素最大限量以及国际食品法典委员会设定的农药和兽药最大残留限量。

6 卫生要求

建议本标准所涉及的产品应遵循《食品卫生总则》（CXC 1-1969）、《乳及乳制品卫生操作规范》（CXC 57-2004）以及《卫生操作规范》和《生产操作规范》等其他相关法典文本。产品应符合《食品微生物标准制定与实施原则和准则》（CXG 21-1997）规定的所有微生物标准。

7 标识

除符合《预包装食品标识通用标准》（CXS 1-1985）和《乳品术语》（CXS 206-1999）外，还应符合以下具体规定。

7.1 产品名称

7.1.1 产品应参考本标准第7.1.3条，酌情根据第2条说明命名。"预包装液态奶油"可以标为"稀奶油"，"加压包装稀奶油"可以标为其他术语形容，体现其特点、用途，或标为"搅打的稀奶油"。不应标为"预制稀奶油"。

本标准涵盖的产品可使用产品生产国和/或产品销售国法律规定的其他名称命名，或以现有通用名称命名，但此类名称在产品销售国不应使消费者对产品特性产生误解。此外，发酵稀奶油产品名称和含量声明等标识可以酌情标注"嗜酸""开菲尔""马奶酒"，但产品应以《发酵乳制品》（CXS 243-2003）第2.1条规定的相应特定发酵剂发酵，并符合《发酵乳标准》第3.3条规定的适用于相应发酵乳的微生物成分标准。

7.1.2 应在名称中或在相同视野醒目位置上，以数值或适当的形容词标注产品销售国可以接受的脂肪含量。

如用营养声明，则应遵循《营养和保健声明使用准则》（CXG 23-1997）。为此，仅以30%的乳脂含量为参考。

7.1.3 采用符合第2.2条和第2.3条说明的乳配料调配或复原生产的稀奶油应标注为"调配稀奶油""复原稀奶油"或以其他形容词如实标注，以免消费者因无此类标识而受误导。

7.1.4 应在名称中或在相同视野醒目位置上适当说明热处理工艺，以免消费者因无此类标识而受误导。

如标识标注所用热处理类型，则应采用食品法典委员会规定的定义。

7.2 乳脂含量说明

乳脂含量应以销售国可接受的方式向最终消费者声明：（i）质量百分比或体积百分比形式；（ii）如果产品标识上标明了份数，也可用每份中乳脂重量（g）表示。

如根据第7.1.2条规定以数值标注产品脂肪含量，则此类标识可视作脂肪含量声明，但标识必须含有以上规定标注的一切补充信息。

7.3 非零售包装标识

在包装容器上除要标明产品名称、批次、生产厂家或包装商的名称和地址外，在包装容器上或附随说明书中，也应按本标准第7条和《预包装食品标识通用标准》（CXS 1-1985）第4.1~4.8条所要求的信息以及储存说明（在必要的地方）加以陈述。如果批次、生产厂家或包装商的名称和地址可以在附随的文件标明，则可以用一个识别标识来代替。

8 抽样和分析方法

为检查产品是否符合本标准，应采用《分析和抽样推荐性方法》（CXS 234-1999）中涉及本标准规定的分析和抽样方法。

STANDARD FOR CREAM AND PREPARED CREAMS

CXS 288-1976

Formerly CODEX STAN A-9-1976. Adopted in 1976. Revised in 2003, 2008.

Amended in 2010, 2018.

1　**SCOPE**

This Standard applies to cream and prepared creams for direct consumption or further processing as defined in Section 2 of this Standard.

2　**DESCRIPTION**

2.1　***Cream*** is the fluid[1] milk product comparatively rich in fat, in the form of an emulsion of fat-in-skimmed milk, obtained by physical separation from milk.

2.2　***Reconstituted cream*** is cream obtained by reconstituting milk products with or without the addition of potable water and with the same end product characteristics as the product described in Section 2.1.

2.3　***Recombined cream*** is cream obtained by recombining milk products with or without the addition of potable water and with the same end product characteristics as the product described in Section 2.1.

2.4　***Prepared creams*** are the milk products obtained by subjecting cream, reconstituted cream and/or recombined cream to suitable treatments and processes to obtain the characteristic properties as specified below.

2.4.1　***Prepackaged liquid cream*** is the fluid[1] milk product obtained by preparing and packaging cream, reconstituted cream and/or recombined cream for direct consumption and/or for direct use as such.

2.4.2　***Whipping cream*** is the fluid[1] cream, reconstituted cream and/or recombined cream that is intended for whipping. When cream is intended for use by the final consumer the cream should have been prepared in a way that facilitates the whipping process.

2.4.3　***Cream packed under pressure*** is the fluid[1] cream, reconstituted cream and/or recombined cream that is packed with a propellant gas in a pressure-propulsion container and which becomes Whipped Cream when removed from that container.

2.4.4　***Whipped cream*** is the fluid[1] cream, reconstituted cream and/or recombined cream into which air or inert gas has been incorporated without reversing the fat-in-skimmed milk emulsion.

2.4.5　***Fermented cream*** is the milk product obtained by fermentation of cream, reconstituted cream or recombined cream, by the action of suitable micro-organisms, that results in reduction of pH with or without coagulation. Where the content of (a) specific micro-organism(s) is(are) indicated, directly or indirectly, in the labelling or otherwise indicated by content claims in connection with sale, these shall be present, viable, active and abun-

---

1　Fluid means capable of pouring at temperatures above freezing.

dant in the product to the date of minimum durability. If the product is heat-treated after fermentation the requirement for viable micro-organisms does not apply.

2.4.6 **Acidified cream** is the milk product obtained by acidifying cream, reconstituted cream and/or recombined cream by the action of acids and/or acidity regulators to achieve a reduction of pH with or without coagulation.

3 ESSENTIAL COMPOSITION AND QUALITY FACTORS

3.1 Raw materials

All creams and prepared creams:

Milk, which may have been subjected to mechanical and physical treatments prior to cream processing.

Additionally, for creams made by reconstitution or recombination:

Butter[1], milk fat products[1], milk powders[1], cream powders[1], and potable water

Additionally, for prepared creams described in Section 2.4.2 through to Section 2.4.6:

The product that remains after the removal of milk fat by churning milk and cream to manufacture butter and milk fat products (often referred to as buttermilk) and that may have been concentrated and/or dried.

3.2 Permitted ingredients

Only those ingredients listed below may be used for the purposes and product categories specified, and only within the limitations specified.

For use in products only for which stabilizers and/or thickeners are justified (see table in Section 4):

- Products derived exclusively from milk or whey and containing 35% (m/m) or more of milk protein of any type (including casein and whey protein products and concentrates and any combinations thereof) and milk powders: These products can be used in the same function as thickeners and stabilizers, provided they are added only in amounts functionally necessary not exceeding 20 g/kg, taking into account any use of the stabilizers and thickeners listed in Section 4;
- Gelatine and starch: These substances can be used in the same function as stabilizers, provided they are added only in amounts functionally necessary as governed by Good Manufacturing Practice taking into account any use of the stabilizers/thickeners listed in section 4.

Additionally, for use in fermented cream, only:

- Starter cultures of harmless micro-organisms including those specified in Section 2 of the *Standard for Fermented Milks* (CXS 243-2003).

Additionally, for use in fermented cream and acidified cream, only:

- Rennet and other safe and suitable coagulating enzymes to improve texture without achieving enzymatic coagulation;
- Sodium chloride.

---

1 For specification, see relevant Codex Standard.

3.3 Composition

Milk fat: Minimum 10% (w/w).

Compositional modification below the minimum specified above for milk fat is not considered to be in compliance with the Section 4.3.3 of the *General Standard for the Use of Dairy Terms* (CXS 206-1999).

4 FOOD ADDITIVES

Only those additives classes indicated in the table below may be used for the product categories specified. Within each additive class, and where permitted according to the table, only those additives listed below may be used and only within the limits specified.

Stabilizers and thickeners, including modified starches may be used singly or in combination, in compliance with the definitions for milk products and only to the extent that they are functionally necessary, taking into account any use of gelatine and starch as provided for in Section 3.2.

Product category	Additive functional class			
	Stabilizers[a]	Acidity regulators[a]	Thickeners[a] and emulsifiers[a]	Packing gases and propellants
Prepackaged liquid cream (2.4.1)	×	×	×	−
Whipping cream (2.4.2)	×	×	×	−
Cream packed under pressure (2.4.3)	×	×	×	×
Whipped cream (2.4.4)	×	×	×	×
Fermented cream (2.4.5)	×	×	×	−
Acidified cream (2.4.6)	×	×	×	−

(a) These additives may be used when needed to ensure product stability and integrity of the emulsion, taking into consideration the fat content and durability of the product. With regard to the durability, special consideration should be given to the level of heat treatment applied since some minimally pasteurized products do not require the use of certain additives.

× The use of additives belonging to the class is technologically justified.

− The use of additives belonging to the class is not technologically justified.

INS no.	Name of additive	Maximum level
Acidity Regulators		
270	Lactic acid, *L*-, *D*- and *DL*-	Limited by GMP
325	Sodium lactate	
326	Potassium lactate	

INS no.	Name of additive	Maximum level
327	Calcium lactate	Limited by GMP
330	Citric acid	
333	Calcium citrates	
500(i)	Sodium carbonate	
500(ii)	Sodium hydrogen carbonate	
500(iii)	Sodium sesquicarbonate	
501(i)	Potassium carbonate	
501(ii)	Potassium hydrogen carbonate	
Stabilizers and thickeners		
170(i)	Calcium carbonate	Limited by GMP
331(i)	Sodium dihydrogen citrate	
331(iii)	Trisodium citrate	
332(i)	Potassium dihydrogen citrate	
332(ii)	Tripotassium citrate	
516	Calcium sulphate	
339(i)	Monosodium dihydrogen phosphate	1 100 mg/kg expressed as phosphorus
339(ii)	Disodium hydrogen phosphate	
339(iii)	Trisodium phosphate	
340(i)	Potassium dihydrogen phophate	
340(ii)	Dipotassium hydrogen phosphate	
340(iii)	Tripotassium phosphate	
341(i)	Calcium dihydrogen phophate	
341(ii)	Calcium hydrogen phosphate	
341(iii)	Tricalcium phosphate	
450(i)	Disodium diphosphate	
450(ii)	Trisodium diphosphate	
450(iii)	Tetrasodium diphosphate	
450(v)	Tetrapotassium diphosphate	
450(vi)	Calcium diphosphate	
450(vii)	Calcium dihydrogen diphosphate	
451(i)	Pentasodium triphosphate	
451(ii)	Pentapotassium triphosphate	
452(i)	Sodium polyphosphate	
452(ii)	Potassium polyphosphate	

INS no.	Name of additive	Maximum level
452(iii)	Sodium calcium polyphosphate	1 100 mg/kg expressed as phosphorus
452(iv)	Calcium polyphosphate	
452(v)	Ammonium polyphosphate	
400	Alginic acid	Limited by GMP
401	Sodium alginate	
402	Potassium alginate	
403	Ammonium alginate	
404	Calcium alginate	
405	Propylene glycol alginate	5 000 mg/kg
406	Agar	Limited by GMP
407	Carrageenan	
407a	Processed euchema seaweed (PES)	
410	Carob bean gum	
412	Guar gum	
414	Gum arabic (Acacia gum)	
415	Xanthan gum	
418	Gellan gum	
440	Pectins	
460(i)	Microcrystalline cellulose (Cellulose gel)	
460(ii)	Powdered cellulose	
461	Methyl cellulose	
463	Hydroxypropyl cellulose	
464	Hydroxypropyl methyl cellulose	
465	Methyl ethyl cellulose	
466	Sodium carboxymethyl cellulose (Cellulose gum)	
472e	Diacetyltartaric and fatty acid esters of glycerol	5 000 mg/kg
508	Potassium chloride	Limited by GMP
509	Calcium chloride	
1410	Monostarch phosphate	
1412	Distarch phosphate	
1413	Phosphated distarch phosphate	
1414	Acetylated distarch phosphate	
1420	Starch acetate	
1422	Acetylated distarch adipate	

INS no.	Name of additive	Maximum level
1440	Hydroxypropyl starch	Limited by GMP
1442	Hydroxypropyl distarch phosphate	
1450	Starch sodium octenyl succinate	
Emulsifiers		
322(i)	Lecithin	Limited by GMP
432	Polyixyethylene (20) sorbitan monolaurate	1 000 mg/kg
433	Polyixyethylene (20) sorbitan monooleate	
434	Polyixyethylene (20) sorbitan monopalmitate	
435	Polyixyethylene (20) sorbitan monostearate	
436	Polyixyethylene (20) sorbitan tristearate	
471	Mono- and diglycerides of fatty acids	Limited by GMP
472a	Acetic and fatty acid esters of glycerol	
472b	Lactic and fatty acid esters of glycerol	
472c	Citric and fatty acid esters of glycerol	
473	Sucrose esters of fatty acids	5 000 mg/kg
475	Polyglycerol esters of fatty acids	6 000 mg/kg
491	Sorbitan monostearate	5 000 mg/kg
492	Sorbitan tristearate	
493	Sorbitan monolaurate	
494	Sorbitan monooleate	
495	Sorbitan monopalmitate	
Packing gases		
290	Carbon dioxide	Limited by GMP
941	Nitrogen	
Propellant		
942	Nitrous oxide	Limited by GMP

5 **CONTAMINANTS**

The products covered by this Standard shall comply with the maximum levels for contaminants that are specified for the product in the *General Standard for Contaminants and Toxins in Food and Feed* (CXS 193-1995).

The milk used in the manufacture of the products covered by this Standard shall comply with the maximum levels for contaminants and toxins specified for milk by the *General Standard for Contaminants and Toxins in Food and Feed* (CXS 193-1995) and with the maximum residue limits for veterinary drug residues and pesticides established for milk by the CAC.

6 HYGIENE

It is recommended that the products covered by the provisions of this standard be prepared and handled in accordance with the appropriate sections of the *General Principles of Food Hygiene* (CXC 1-1969), the *Code of Hygienic Practice for Milk and Milk Products* (CXC 57-2004) and other relevant Codex texts such as Codes of Hygienic Practice and Codes of Practice. The products should comply with any microbiological criteria established in accordance with the *Principles and Guidelines for the Establishment and Application of Microbiological Criteria Related to Foods* (CXG 21-1997).

7 LABELLING

In addition to the provisions of the *General Standard for the Labelling of Prepackaged Foods* (CXS 1-1985) and the *General Standard for the Use of Dairy Terms* (CXS 206-1999), the following specific provisions apply.

7.1 Name of the food

7.1.1 The name of the food shall be as specified in Section 2 of this Standard, as appropriate and taking into account section 7.1.3. However, "prepackaged liquid cream" may be designated as "cream" and "cream packed under pressure" may be designated by another descriptive term that refers to its nature or intended use or as "Whipped Cream". The term "prepared cream" should not apply as a designation.

The products covered by this Standard may alternatively be designated with other names specified in the national legislation of the country in which the product is manufactured and/or sold or with a name existing by common usage, provided that such designations do not create an erroneous impression in the country of retail sale regarding the character and identity of the food.

In addition, labelling statements, such as product designation of fermented creams and content claims, may include reference to the terms "Acidophilus", "Kefir", and "Kumys", as appropriate, provided that the product has been fermented by the corresponding specific starter culture(s) specified in Section 2.1 of the *Standard for Fermented Milks* (CXS 243-2003), and provided that the product complies with those compositional microbiological criteria that are applicable to the corresponding fermented milk as specified in Section 3.3 of that Standard.

7.1.2 The designation shall be accompanied by an indication of the fat content that is acceptable in the country of retail sale, either as a numerical value or by a suitable qualifying term, either as part of the name or in a prominent position in the same field of vision.

Nutrition claims, when used, shall be in accordance with the *Guidelines for Use of Nutrition and Health Claims* (CXG 23-1997). For this purpose only, the level of 30% milk fat constitutes the reference.

7.1.3 Creams which have been manufactured by the recombination or reconstitution of dairy ingredients as specified in Sections 2.2 and 2.3 shall be labelled as "Recombined cream" or "Reconstituted cream" or another truthful qualifying term if the consumer would be misled by the absence of such labelling.

7.1.4 An appropriate description of the heat treatment should be given, either as part of the name or in a prominent position in the same field of vision, if the consumer would be misled by the absence of such labelling.

When reference is made in the labelling to the type of heat treatment(s) applied, the definitions established by the Codex Alimentarius Commission shall apply.

7.2 Declaration of milk fat content

The milk fat content shall be declared in a manner acceptable in the country of sale to the final consumer, either as (i) a percentage of mass or volume, (ii) in grams per serving as qualified in the label, provided that the number of servings is stated.

Where the fat content of the product is indicated by a numerical value in accordance with Section 7.1.2, such indication may constitute the fat declaration, provided that the indication includes any additional information as required above.

7.3 Labelling of non-retail containers

Information required in Section 7 of this Standard and Sections 4.1 to 4.8 of the *General Standard for the Labelling of Prepackaged Foods* (CXS 1-1985), and, if necessary, storage instructions, shall be given either on the container or in accompanying documents, except that the name of the product, lot identification, and the name and address of the manufacturer or packer shall appear on the container. However, lot identification and the name and address of the manufacturer or packer may be replaced by an identification mark provided that such a mark is clearly identifiable with the accompanying documents.

8 METHODS OF SAMPLING AND ANALYSIS

For checking the compliance with this Standard, the methods of analysis and sampling contained in the *Recommended Methods of Analysis and Sampling* (CXS 234-1999) relevant to the provisions in this Standard, shall be used.

乳 清 粉

STANDARD FOR WHEY POWDERS

CXS 289-1995

前为CODEX STAN A-15-1995。1995年通过。

2003年修订。2006年、2010年、2018年修正。

1　范围

本标准适用于符合本标准第2条所述即食或用于进一步加工的乳清粉和酸乳清粉。

2　说明

乳清粉是乳清或酸乳清干燥而成的乳制品。

乳清是在干酪、酪蛋白或类似产品生产过程中，乳和/或乳制品凝固成凝乳后分离而成的液态乳制品。凝固主要通过凝乳酶类作用形成。

酸乳清是在干酪、酪蛋白或类似产品生产过程中，乳和/或乳制品凝固成凝乳后分离而成的液态乳制品。凝固主要通过酸化形成。

3　基本成分和质量要求

3.1　基本原料

乳清或酸乳清。

3.2　其他配料

在预结晶乳清粉生产过程中使用种子乳糖[1]。

3.3　成分

乳清粉：

标准	最低含量	参考含量	最高含量
乳糖[a]	/	61.0%（m/m）	/
乳蛋白[b]	10.0%（m/m）	/	/
乳脂	/	2.0%（m/m）	/
水[c]	/	/	5.0%（m/m）
灰分	/	/	9.5%（m/m）
pH值（在10%溶液中）[d]	>5.1	/	/

---

[1]　见《糖类标准》（CXS 212-1999）。

酸乳清粉：

标准	最低含量	参考含量	最高含量
乳糖(a)	/	61.0%（m/m）	/
乳蛋白(b)	7.0%（m/m）	/	/
乳脂	/	2.0%（m/m）	/
水(c)	/	/	4.5%（m/m）
灰分	/	/	15.0%（m/m）
pH值（在10%溶液中）(e)	/	/	5.1

（a）产品可能同时含有无水乳糖和一水乳糖，但乳糖含量仍以无水乳糖表示。每100份一水乳糖含95份无水乳糖。
（b）蛋白质含量按测定的总凯氏氮乘以6.38计算。
（c）水分含量不包括乳糖结晶水。
（d）或可滴定酸度（以乳酸计）<0.35%。
（e）或可滴定酸度（以乳酸计）≥0.35%。

根据《乳品术语》（CXS 206-1999）第4.3.3条规定，可以调整乳清粉成分，满足最终产品理想的成分需求，例如用于中和或除盐。然而，成分调整超过上述乳蛋白和水分最小或最大限量，视为不符合第4.3.3条规定。

4 食品添加剂

可以根据本标准在食品中使用《食品添加剂通用标准》（CXS 192-1995）表1和表2食品类别01.8.2（乳清粉和乳清制品，不包括乳清干酪）所列食品添加剂。

5 污染物

本标准所涉及的产品应符合《食品及饲料中污染物和毒素通用标准》（CXS 193-1995）规定的污染物最大限量。

本标准所涉及的产品在加工中使用的牛乳应符合《食品及饲料中污染物和毒素通用标准》（CXS 193-1995）规定的乳中污染物和毒素最大限量以及国际食品法典委员会设定的农药和兽药最大残留限量。

6 卫生要求

建议本标准所涉及的产品应遵循《食品卫生总则》（CXC 1-1969）、《乳及乳制品卫生操作规范》（CXC 57-2004）以及《卫生操作规范》和《生产操作规范》等其他相关法典文本。产品应符合《食品微生物标准制定与实施原则和准则》（CXG 21-1997）规定的所有微生物标准。

7 标识

除符合《预包装食品标识通用标准》（CXS 1-1985）和《乳品术语》（CXS 206-1999）外，还应符合以下具体规定。

7.1 产品名称

产品名称应为：

乳清粉	参照第2条所做定义和第3.3条所述成分
酸乳清粉	

对于脂肪和/或乳糖含量超出本标准第3.3条规定参考含量的产品，名称应带有适当修饰语作为名称一部分或标注于相同视野显眼位置，说明所做调整或乳糖和/或脂肪含量。

如乳清粉达到下列成分标准，乳清粉名称可以加注"甜"字样：

乳糖最低含量：	65%
蛋白质最低含量：	11%
灰分最高含量：	8.5%
pH值（10%的溶液）*	>6

* 或最高0.16%的可滴定酸度（以乳酸计）。

7.2 非零售包装标识

在包装容器上除要标明产品名称、批次、生产厂家或包装商的名称和地址外，在包装容器上或附随说明书中，也应按本标准第7条和《预包装食品标识通用标准》（CXS 1-1985）第4.1~4.8条所要求的信息以及储存说明（在必要的地方）加以陈述。如果批次、生产厂家或包装商的名称和地址可以在附随的文件标明，则可以用一个识别标识来代替。

8 抽样和分析方法

为检查产品是否符合本标准，应采用《分析和抽样推荐性方法》（CXS 234-1999）中涉及本标准规定的分析和抽样方法。

STANDARD FOR WHEY POWDERS

CXS 289-1995

Formerly CODEX STAN A-15-1995. Adopted in 1995. Revised in 2003.
Amended in 2006, 2010, 2018.

1 SCOPE

This Standard applies to Whey Powder and Acid Whey Powder, intended for direct consumption or further processing, in conformity with the description in Section 2 of this Standard.

2 DESCRIPTION

Whey powders are milk products obtained by drying whey or acid whey.

Whey is the fluid milk product obtained during the manufacture of cheese, casein or similar products by separation from the curd after coagulation of milk and/or of products obtained from milk. Coagulation is obtained through the action of, principally, rennet type enzymes.

Acid whey is the fluid milk product obtained during the manufacture of cheese, casein or similar products by separation from the curd after coagulation of milk and/or of products obtained from milk. Coagulation is obtained, principally, by acidification.

3 ESSENTIAL COMPOSITION AND QUALITY FACTORS

3.1 Raw materials

Whey or acid whey.

3.2 Permitted ingredients

Seed lactose[1] in the manufacture of pre-crystallized whey powder.

3.3 Composition

Whey powder:

Criteria	Minimum content	Reference content	Maximum content
Lactose[a]	n.s.	61.0% (m/m)	n.s.
Milk protein[b]	10.0% (m/m)	n.s.	n.s.
Milk fat	n.s.	2.0% (m/m)	n.s.
Water[c]	n.s.	n.s.	5.0% (m/m)
Ash	n.s	n.s.	9.5% (m/m)
pH (in 10% solution)[d]	> 5.1	n.s.	n.s.

---

1 See *Standard for Sugars* (CXS 212-1999).

Acid whey powder:

Criteria	Minimum content	Reference content	Maximum content
Lactose[a]	n.s.	61.0% (m/m)	n.s.
Milk protein[b]	7.0% (m/m)	n.s.	n.s.
Milk fat	n.s.	2.0% (m/m)	n.s.
Water[c]	n.s.	n.s.	4.5% (m/m)
Ash	n.s.	n.s.	15.0% (m/m)
pH (in 10% solution)[e]	n.s.	n.s.	5.1

(a) Although the products may contain both anhydrous lactose and lactose monohydrate, the lactose content is expressed as anhydrous lactose. 100 parts of lactose monohydrate contain 95 parts of anhydrous lactose.

(b) Protein content is 6.38 multiplied by the total Kjeldahl nitrogen determined.

(c) The water content does not include water of crystallization of the lactose.

(d) Or titratable acidity (calculated as lactic acid) <0.35%.

(e) Or titratable acidity (calculated as lactic acid) ⩾0.35%.

In accordance with the provision of Section 4.3.3 of the *General Standard for the Use of Dairy Terms* (CXS 206-1999), whey powders may be modified in composition to meet the desired end-product composition, for instance, neutralization or demineralization. However, compositional modifications beyond the minima or maxima specified above for milk protein and water are not considered to be incompliance with the Section 4.3.3.

4 FOOD ADDITIVES

Food additives listed in Tables 1 and 2 of the *General Standard for Food Additives* (CXS 192-1995) in Food Category 01.8.2 (Dried whey and whey products, excluding whey cheese) may be used in foods subject to this standard.

5 CONTAMINANTS

The products covered by this Standard shall comply with the maximum levels for contaminants that are specified for the product in the General *Standard for Contaminants and Toxins in Food and Feed* (CXS 193-1995).

The milk used in the manufacture of the products covered by this Standard shall comply with the maximum levels for contaminants and toxins specified for milk by the *General Standard for Contaminants and Toxins in Food and Feed* (CXS 193-1995) and with the maximum residue limits for veterinary drug residues and pesticides established for milk by the CAC.

6 HYGIENE

It is recommended that the products covered by the provisions of this standard be prepared and handled in accordance with the appropriate sections of the *General Principles of Food Hygiene* (CXC 1-1969), the *Code of Hygienic Practice for Milk and Milk Products* (CXC 57-2004) and other relevant Codex texts such as Codes of Hygienic Practice and Codes of Practice. The products should comply with any microbiological criteria established in accordance with the *Principles and Guidelines for the Establishment and Application of Microbiological Criteria Related to Foods* (CXG 21-1997).

7 LABELLING

In addition to the provisions of the General *Standard for the Labelling of Prepackaged Foods* (CXS 1-1985) and the *General Standard for the Use of Dairy Terms* (CXS 206-1999), the following specific provisions apply.

7.1 Name of the food

The name of the food shall be:

Whey powder	According to the definitions in Section 2 and compositions as specified in Section 3.3
Acid whey powder	

The designation of products in which the fat and/or lactose contents are below or above the reference content levels specified in Section 3.3 of this Standard shall be accompanied by an appropriate qualification describing the modification made or the lactose and/or fat content, respectively, either as part of the name or in a prominent position in the same field of vision.

The term "sweet" may accompany the name of whey powder, provided that the whey powder meets the following compositional criteria:

minimum lactose:	65%
minimum protein:	11%
maximum ash:	8.5%
pH (10% solution)[*]:	>6

[*] or titratable acidity of maximum 0.16% (calculated as lactic acid).

7.2 Labelling of non-retail containers

Information required in Section 7 of this Standard and Sections 4.1 to 4.8 of the *General Standard for the Labelling of Prepackaged Foods* (CXC 1-1985) and, if necessary, storage instructions, shall be given either on the container or in accompanying documents, except that the name of the product, lot identification and the name and address of the manufacturer or packer shall appear on the container. However, lot identification and the name and address of the manufacturer or packer may be replaced by an identification mark, provided that such mark is clearly identifiable with the accompanying documents.

8 METHODS OF SAMPLING AND ANALYSIS

For checking the compliance with this Standard, the methods of analysis and sampling contained in the *Recommended Methods of Analysis and Sampling* (CXS 234-1999) relevant to the provisions in this Standard, shall be used.

食用酪蛋白制品

STANDARD FOR EDIBLE CASEIN PRODUCTS

CXS 290-1995

原为CODEX STAN A-18-1995。1995年通过。2001年修订。

2010年、2013年、2014年、2016年、2018年修正。

1 范围

本标准适用于符合本标准第2条所述即食或用于进一步加工的食用酸性酪蛋白、食用凝乳酶酪蛋白和食用酪蛋白酸盐。

2 说明

食用酸性酪蛋白指对脱脂乳和/或其他乳制品经酸沉淀形成的凝块进行分离、洗涤和干燥而获得的乳制品。

食用凝乳酶酪蛋白指对脱脂乳和/或其他乳制品的凝块进行分离、洗涤和干燥而获得的乳制品。其中的凝块通过凝乳酶或其他凝固酶的作用获得。

食用酪蛋白酸盐指使用中和剂对食用酪蛋白或食用酪蛋白凝块进行处理后再干燥而获得的乳制品。

3 基本成分和质量要求

3.1 基本原料

脱脂乳和/或其他乳制品。

3.2 其他配料

- 无害乳酸发酵剂；
- 凝乳酶或其他安全、适宜的凝固酶；
- 饮用水。

3.3 成分

	凝乳酶酪蛋白	酸性酪蛋白	酪蛋白酸盐
干物质中乳蛋白最低含量[a]	84.0% m/m	90.0% m/m	88.0% m/m
乳蛋白中酪蛋白最低含量	95.0% m/m	95.0% m/m	95.0% m/m
水分最高含量[b]	12.0% m/m	12.0% m/m	8.0% m/m
乳脂最高含量	2.0% m/m	2.0% m/m	2.0% m/m

	凝乳酶酪蛋白	酸性酪蛋白	酪蛋白酸盐
灰分（包括P_2O_5）	≥7.5% m/m	≤2.5% m/m	–
乳糖最高含量(c)	1.0% m/m	1.0% m/m	1.0% m/m
游离酸最高含量	–	0.27 mL 0.1 N NaOH/g	–
最大pH值	–	–	8.0

（a）蛋白质含量等于6.38乘以凯氏法测定的总氮值。
（b）水分含量不包括乳糖中结晶水。
（c）产品中可能既有无水乳糖又有一水乳糖，但乳糖含量应以无水乳糖表示。100份一水乳糖中含95份无水乳糖。

按照《乳品术语》（CXS 206-1999）第4.3.3条规定，食用酪蛋白制品可以根据最终产品需要调整成分。但成分调整如果超出上述规定干物质中乳蛋白、酪蛋白、水分、乳脂、乳糖和游离酸含量的最大值或最小值，则视为违反第4.3.3条规定。

4　食品添加剂

只允许在规定范围内使用下表列出的食品添加剂。

酪蛋白酸盐

INS编号	添加剂名称	最大限量
酸度调节剂		
170	碳酸钙	根据GMP限量使用
261（i）	醋酸钾	
262（i）	乙酸钠	
263	乙酸钙	
325	乳酸钠	
326	乳酸钾	
327	乳酸钙	
329	DL-乳酸镁	
331	柠檬酸钠	
332	柠檬酸钾	
333	柠檬酸钙	
345	柠檬酸镁	
380	柠檬酸三铵	
339	磷酸钠	4 400 mg/kg，单用或混用，以磷计*
340	磷酸钾	

INS编号	添加剂名称	最大限量
341	磷酸钙	4 400 mg/kg，单用或混用，以磷计*
342	磷酸铵	
343	磷酸镁	
452	聚磷酸盐	2 200 mg/kg，单用或混用，以磷计*
500	碳酸钠	根据GMP限量使用
501	碳酸钾	
503	碳酸铵	
504	碳酸镁	
524	氢氧化钠	
525	氢氧化钾	
526	氢氧化钙	
527	氢氧化铵	
528	氢氧化镁	
乳化剂		
322	卵磷脂	根据GMP限量使用
471	单、双甘油脂肪酸酯	
疏松剂		
325	乳酸钠	根据GMP限量使用
抗结块剂		
170（i）	碳酸钙	4 400 mg/kg，单用或混用
341（iii）	磷酸三钙	
343（iii）	磷酸三镁	
460	纤维素	
504（i）	碳酸镁	
530	氧化镁	
551	二氧化硅（无定型）	
552	硅酸钙	
553	硅酸镁	
554	硅酸铝钠	265 mg/kg，以铝计
1422	乙酰化双淀粉己二酸酯	4 400 mg/kg，单用或混用

* 磷总量不得超过4 400 mg/kg。

5 污染物

本标准所涉及的产品应符合《食品及饲料中污染物和毒素通用标准》（CXS 193-1995）规定的污染物最大限量。

本标准所涉及的产品在加工中使用的牛乳应符合《食品及饲料中污染物和毒素通用标准》（CXS 193-1995）规定的乳中污染物和毒素最大限量以及国际食品法典委员会设定的农药和兽药最大残留限量。

6 卫生要求

建议本标准所涉及的产品应遵循《食品卫生总则》（CXC 1-1969）、《乳及乳制品卫生操作规范》（CXC 57-2004）以及《卫生操作规范》和《生产操作规范》等其他相关法典文本。产品应符合《食品微生物标准制定与实施原则和准则》（CXG 21-1997）规定的所有微生物标准。

7 标识

除符合《预包装食品标识通用标准》（CXS 1-1985）和《乳品术语》（CXS 206-1999）外，还应符合下列具体标准。

7.1 产品名称

产品名称应为：

- 食用酸性酪蛋白
- 食用酪蛋白酸盐
- 食用凝乳酶酪蛋白

按照第2条中的说明和第3.3条中的成分

食用酪蛋白酸盐的名称须标明所用阳离子。

7.2 非零售包装标识

在包装容器上除要标明产品名称、批次、生产厂家或包装商的名称和地址外，在包装容器上或附随说明书中，也应按本标准第7条和《预包装食品标识通用标准》（CXS 1-1985）第4.1～4.8条所要求的信息以及储存说明（在必要的地方）加以陈述。如果批次、生产厂家或包装商的名称和地址可以在附随的文件标明，则可以用一个识别标识来代替。

8 抽样和分析方法

为检查产品是否符合本标准，应采用《分析和抽样推荐性方法》（CXS 234-1999）中涉及本标准规定的分析和抽样方法。

附录

附加信息

以下补充信息对前文各条款的规定不构成影响,前文规定内容对产品标识、食品名称的使用以及食品安全性至关重要。

1 **其他质量指标**

1.1 **外观**

呈纯白至浅奶油色;无结块,即使有也轻压即碎。

1.2 **味道和气味**

产品只有轻微其他味道和气味,不含令人不快的味道和气味。

2 **食品添加剂**

用于沉淀的酸类:

INS编号	添加剂名称
260	冰醋酸
270	乳酸（*L*-,*D*-和*DL*-）
330	柠檬酸
338	磷酸
507	盐酸
513	硫酸
用于促进凝乳过程	
509	氯化钙

3 **附加质量指标**

最大沉淀量	凝乳酶酪蛋白	酸性酪蛋白	酪蛋白酸盐
焦糊颗粒	15 mg/25 g	22.5 mg/25 g	22.5 mg/25 g（喷雾干燥） 81.5 mg/25 g（滚筒干燥）

重金属

以下限量适用:

金属	最大限量
铜	5 mg/kg
铁	20 mg/kg（滚筒干燥的酪蛋白酸盐为50 mg/kg）

4 **其他分析方法**

为检查产品是否符合本标准,应采用《分析和抽样推荐性方法》（CXS 234-1999）中涉及本标准规定的分析和抽样方法。

STANDARD FOR EDIBLE CASEIN PRODUCTS

CXS 290-1995

Formerly CODEX STAN A-18-1995. Adopted in 1995. Revised in 2001.

Amended in 2010, 2013, 2014, 2016, 2018.

1 SCOPE

This Standard applies to edible acid casein, edible rennet casein and edible caseinate, intended for direct consumption or further processing, in conformity with the description in Section 2 of this Standard.

2 DESCRIPTION

Edible acid casein is the milk product obtained by separating, washing and drying the acid-precipitated coagulum of skimmed milk and/or of other products obtained from milk.

Edible rennet casein is the milk product obtained by separating, washing and drying the coagulum of skimmed milk and/or of other products obtained from milk. The coagulum is obtained through the reaction of rennet or other coagulating enzymes.

Edible caseinate is the milk product obtained by action of edible casein or edible casein curd coagulum with neutralizing agents followed by drying.

3 ESSENTIAL COMPOSITION AND QUALITY FACTORS

3.1 Raw materials

Skimmed milk and/or other products obtained from milk.

3.2 Permitted ingredients

- Starter cultures of harmless lactic acid producing bacteria;
- Rennet or other safe and suitable coagulating enzymes;
- Potable water.

3.3 Composition

	Rennet casein	Acid casein	Caseinates
Minimum milk protein in dry matter[a]	84.0% m/m	90.0% m/m	88.0% m/m
Minimum content of casein in milk protein	95.0% m/m	95.0% m/m	95.0% m/m
Maximum water[b]	12.0% m/m	12.0% m/m	8.0% m/m
Maximum milkfat	2.0% m/m	2.0% m/m	2.0% m/m
Ash (including P_2O_5)	7.5% m/m (min.)	2.5% m/m (max.)	–
Maximum lactose[c]	1.0% m/m	1.0% m/m	1.0% m/m

	Rennet casein	Acid casein	Caseinates
Maximum free acid	–	0.27 mL 0.1 N NaOH/g	–
Maximum pH value	–	–	8.0

(a) Protein content is 6.38 multiplied by the total Kjeldahl nitrogen determined.
(b) The water content does not include water of crystallization of the lactose.
(c) Although the products may contain both anhydrous lactose and lactose monohydrate, the lactose content is expressed as anhydrous lactose. 100 parts of lactose monohydrate contain 95 parts of anhydrous lactose.

In accordance with the provision of section 4.3.3 of the *General Standard for the Use of Dairy Terms* (CXS 206-1999), edible casein products may be modified in composition to meet the desired end-product composition. However, compositional modifications beyond the minima or maxima specified above for milk protein in dry matter, casein, water, milkfat, lactose and free acid are not considered to be in compliance with the Section 4.3.3.

4   FOOD ADDITIVES

Only those additives listed below may be used within the limits specified.

Caseinates

INS no.	Name of additive	Maximum level
Acidity regulators		
170	Calcium citrates	Limited by GMP
261(i)	Potassium acetate	
262(i)	Sodium acetate	
263	Calcium acetate	
325	Sodium lactate	
326	Potassium lactate	
327	Calcium lactate	
329	Magnesium lactate, *DL*-	
331	Sodium citrates	
332	Potassium citrates	
333	Calcium citrates	
345	Magnesium citrates	
380	Triammonium citrates	
339	Sodium phosphates	4 400 mg/kg singly or in combination expressed as phosphorous*
340	Potassium phosphates	
341	Calcium phosphates	

INS no.	Name of additive	Maximum level
342	Ammonium phosphates	4 400 mg/kg singly or in combination expressed as phosphorous*
343	Magnesium phosphates	
452	Polyphosphates	2 200 mg/kg singly or in combination expressed as phosphorous*
500	Sodium carbonates	Limited by GMP
501	Potassium carbonates	
503	Ammonium carbonates	
504	Magnesium carbonates	
524	Sodium hydroxide	
525	Potassium hydroxide	
526	Calcium hydroxide	
527	Ammonium hydroxide	
528	Magnesium hydroxide	
Emulsifiers		
322	Lecithins	Limited by GMP
471	Mono- and di-glycerides of fatty acids	
Bulking agents		
325	Sodium lactate	Limited by GMP
Anticaking agents		
170(i)	Calcium carbonate	4 400 mg/kg singly or in combination *
341(iii)	Tricalcium phosphate	
343(iii)	Trimagnesium phosphate	
460	Cellulose	
504(i)	Magnesium carbonate	
530	Magnesium oxide	
551	Silicon dioxide, amorphous	
552	Calcium silicate	
553	Magnesium silicates	
554	Sodium aluminium silicate	265 mg/kg, expressed as aluminium
1442	Hydroxypropyldistach phosphate	4 400 mg/kg singly or in combination *

* Total amount of phosphorous shall not exceed 4 400 mg/kg.

5 CONTAMINANTS

The products covered by this Standard shall comply with the maximum levels for contaminants that are specified for the product in the *General Standard for Contaminants and Toxins in Food and Feed* (CXS 193-1995).

The milk used in the manufacture of the products covered by this Standard shall comply with the maximum levels for contaminants and toxins specified for milk by the *General Standard for Contaminants and Toxins in Food and Feed* (CXS 193-1995) and with the maximum residue limits for veterinary drug residues and pesticides established for milk by the CAC.

6 HYGIENE

It is recommended that the products covered by the provisions of this standard be prepared and handled in accordance with the appropriate sections of the *General Principles of Food Hygiene* (CXC 1-1969), the *Code of Hygienic Practice for Milk and Milk Products* (CXC 57-2004) and other relevant Codex texts such as Codes of Hygienic Practice and Codes of Practice.

The products should comply with any microbiological criteria established in accordance with the *Principles and Guidelines for the Establishment and Application of Microbiological Criteria Related to Foods* (CXG 21-1997).

7 LABELLING

In addition to the provisions of the *General Standard for the Labelling of Prepackaged Foods* (CXS 1-1985) and the *General Standard for the Use of Dairy Terms* (CXS 206-1999), the following specific provisions apply:

7.1 Name of the food

The name of the food shall be:

- Edible acid casein
- Edible caseinate
- Edible rennet casein

According to the descriptions in Section 2 and the compositions in Section 3.3

The name of edible caseinate shall be accompanied by an indication of the cation used.

7.2 Labelling of non-retail containers

Information required in Section 7 of this Standard and Sections 4.1 to 4.8 of the *General Standard for the Labelling of Prepackaged Foods* (CXS 1-1985) and, if necessary, storage instructions, shall be given either on the container or in accompanying documents, except that the name of the product, lot identification and the name and address of the manufacturer or packer shall appear on the container. However, lot identification and the name and address of the manufacturer or packer may be replaced by an identification mark, provided that such mark is clearly identifiable with the accompanying documents.

8 METHODS OF SAMPLING AND ANALYSIS

For checking the compliance with this Standard, the methods of analysis and sampling contained in the *Recommended Methods of Analysis and Sampling* (CXS 234-1999) relevant to the provisions in this Standard, shall be used.

APPENDIX

ADDITIONAL INFORMATION

The additional information below does not affect the provisions in the preceding sections which are those that are essential to the product identity, the use of the name of the food and the safety of the food.

1 OTHER QUALITY FACTORS

1.1 Physical appearance

White to pale cream; free from lumps which do not break up under slight pressure.

1.2 Flavour and odour

Not more than slight foreign flavours and odours. The product must be free from offensive flavours and odours.

2 PROCESSING AIDS

Acids used for precipitation purposes:

INS no.	Name
260	Acetic acid, glacial
270	Lactic acid, *L*-, *D*- and *DL*-
330	Citric acid
338	Orthophosphoric acid
507	Hydrochloric acid
513	Sulphuric acid
For renneting enhancement purposes	
509	Calcium chloride

3 ADDITIONAL QUALITY FACTORS

Maximum sediment	Rennet casein	Acid casein	Caseinates
scorched particles	15 mg/25 g	22.5 mg/25 g	22.5 mg/25 g (spray dried) 81.5 mg/25 g (roller dried)

Heavy metals

The following limits apply:

Metal	Maximum limit
Copper	5 mg/kg
Iron	20 mg/kg (50 mg/kg in roller dried caseinates)

4 ADDITIONAL METHODS OF ANALYSIS

For checking the compliance with this Standard, the methods of analysis and sampling contained in the *Recommended Methods of Analysis and Sampling* (CXS 234-1999) relevant to the provisions in this Standard, shall be used.

乳制品渗透物粉

STANDARD FOR DAIRY PERMEATE POWDERS

CXS 331-2017

2017年通过。

1 范围

本标准适用于符合本标准第2条所述即食或用于进一步加工的乳制品渗透物粉。

2 说明

乳制品渗透物粉是指干燥乳制品[1]，特点是乳糖含量高：

（a）由乳、乳清[2]（酸乳清除外）、稀奶油[3]和/或甜酪乳和/或类似原材料通过使用膜渗透技术尽可能去除乳脂和乳蛋白获得的低蛋白物制成，和/或；

（b）由（a）中所列的同样原材料通过去除乳脂但保留乳糖的其他加工技术获得，最终产品构成与第3.3条中描述相同。

乳清渗透物粉是指由低蛋白乳清加工制成的低蛋白乳粉。低蛋白乳清由乳清去除乳清蛋白但保留乳糖获得。

牛乳渗透物粉是指由低蛋白乳加工制成的低蛋白乳粉[4]。

3 基本成分和质量要求

3.1 基本原料

乳制品渗透物粉：牛乳渗透物、乳清渗透物、奶油渗透物、甜酪乳渗透物和/或；类似的含糖乳制品。

乳清渗透物粉：低蛋白乳清。

牛乳渗透物粉：低蛋白乳。

3.2 其他配料

预结晶产品加工过程中可加入种子乳糖[5]。

---

1 乳制品定义见《乳品术语》（CXS 206-1999）。
2 乳清定义见《乳清粉标准》（CXS 289-1995）。
3 稀奶油定义见《稀奶油和预制稀奶油法典标准》（CXS 288-1976）。
4 低蛋白乳定义见《乳粉和奶油粉法典标准》（CXS 207-1999）。
5 乳糖定义见《糖类标准》（CXS 212-1999）。

3.3 成分

标准	乳制品渗透物粉	乳清渗透物粉	牛乳渗透物粉
无水乳糖最低含量[a]（m/m）	76.0%	76.0%	76.0%
最高氮含量（m/m）	1.1%	1.1%	0.8%
最高乳脂含量（m/m）	1.5%	1.5%	1.5%
最高灰分含量（m/m）	14.0%	12.0%	12.0%
最高含水量[b]（m/m）	5.0%	5.0%	5.0%

（a）尽管产品可能同时含有无水乳糖和一水乳糖，但乳糖含量仍以无水乳糖表示。100份一水乳糖中含有95份无水乳糖。

（b）水分含量不包括乳糖中的结晶水。

根据《乳品术语》（CXS 206-1999）第4.3.3条规定，本标准所涉乳制品渗透物粉可根据最终产品需要调整成分，如部分脱盐。但成分调整如超出上述乳糖、氮、乳脂、灰分和水分的最低和最高含量，则视作不符合《乳品术语》第4.3.3条规定。

4 食品添加剂

4.1 本标准所涉乳制品渗透物粉中不得使用食品添加剂。

4.2 食品添加剂

本标准所涉产品中使用的食品添加剂应符合《食品添加剂使用指南》（CXG 75-2010）。

5 污染物

本标准所涉及的产品应符合《食品及饲料中污染物和毒素通用标准》（CXS 193-1995）规定的污染物最大限量。

本标准所涉及在产品加工中使用的牛乳应符合《食品及饲料中污染物和毒素通用标准》（CXS 193-1995）规定的乳中污染物和毒素最大限量以及国际食品法典委员会设定的农药和兽药最大残留限量。

6 卫生要求

建议本标准规定中所涉及的产品应遵循《食品卫生总则》（CXC 1-1969）、《乳及乳制品卫生操作规范》（CXC 57-2004）以及《卫生操作规范》和《生产操作规范》等其他相关法典文本。产品应符合《食品微生物标准制定与实施原则和准则》（CXG 21-1997）规定的所有微生物标准。

7 标识

除符合《预包装食品标识通用标准》（CXS 1-1985）和《乳品术语》（CXS 206-

1999）规定外，还应符合下列具体规定。

7.1 产品名称

产品名称应为"乳制品渗透物粉"。符合第2条相关描述和第3.3条成分的产品可分别称为牛乳渗透物粉和乳清渗透物粉。在产品销售国，食品名称可酌情以"富含乳糖的脱蛋白_____粉"为补充，横线空白处可根据产品特性酌情填写"乳""乳清""奶"等术语。

7.2 非零售包装标识

在包装容器上除要标明产品名称、批次、生产厂家或包装商的名称和地址外，在包装容器上或附随说明书中，也应按本标准第7条和《预包装食品标识通用标准》（CXS 1-1985）第4.1~4.8条所要求的信息以及储存说明（在必要的地方）加以陈述。如果批次、生产厂家或包装商的名称和地址可以在附随的文件标明，则可以用一个识别标识来代替。

8 抽样和分析方法

为检查产品是否符合本标准，应采用《分析和抽样推荐性方法》（CXS 234-1999）中涉及本标准规定的分析和抽样方法。

STANDARD FOR DAIRY PERMEATE POWDERS

CXS 331-2017

Adopted in 2017.

1 SCOPE

This Standard applies to dairy permeate powders, in conformity with the description in Section 2 of this Standard, intended for further processing and/or as ingredient in other foods.

2 DESCRIPTION

Dairy permeate powders are dried milk products[1] characterized by a high content of lactose:

(a) manufactured from permeates which are obtained by removing, through the use of membrane filtration, and to the extent practical, milk fat and milk protein, but not lactose, from milk, whey[2] (excluding acid whey), cream[3] and/or sweet buttermilk, and/or from similar raw materials, and/or;

(b) obtained by other processing techniques involving removal of milk fat and milk protein, but not lactose, from the same raw materials listed under (a) and resulting in an end-product with the same composition as specified in section 3.3.

Whey permeate powder is the dairy permeate powder manufactured from whey permeate. Whey permeate is obtained by removing whey protein, but not lactose, from whey.

Milk permeate powder is the dairy permeate powder manufactured from milk permeate[4].

3 ESSENTIAL COMPOSITION AND QUALITY FACTORS

3.1 Raw materials

Dairy permeate powders: Milk permeate, whey permeate, cream permeate, sweet buttermilk permeate and/or similar lactose-containing milk products.

Whey permeate powder: Whey permeate.

Milk permeate powder: Milk permeate.

3.2 Permitted ingredients

Seed lactose[5] in the manufacture of pre-crystallized products.

---

1 Definition of *milk product*, see *General Standard for the Use of Dairy Terms* (CXS 206-1999)
2 Definition of *whey*, see *Standard for Whey Powders* (CXS 289-1995)
3 Definition of *cream*, see the *Standard for Cream and Prepared Creams* (CXS 288-1976)
4 Definition of *milk permeate*, see *Standard for Milk Powders and Cream Powder* (CXS 207-1999)
5 Definition of *lactose*, see the *Standard for Sugars* (CXS 212-1999)

3.3 Composition

Criteria	Dairy permeate powder	Whey permeate powder	Milk permeate powder
Minimum lactose, anhydrous[a] (m/m)	76.0%	76.0%	76.0%
Maximum nitrogen (m/m)	1.1%	1.1%	0.8 %
Maximum milk fat (m/m)	1.5%	1.5%	1.5%
Maximum ash (m/m)	14.0%	12.0%	12.0%
Maximum moisture[b] (m/m)	5.0%	5.0%	5.0%

(a) Although the products may contain both anhydrous lactose and lactose monohydrate, the lactose content is expressed as anhydrous lactose. 100 parts of lactose monohydrate contain 95 parts of anhydrous lactose.

(b) The moisture content does not include the water of crystallization of the lactose.

In accordance with the provision of section 4.3.3 of the *General Standard for the Use of Dairy Terms* (CXS 206-1999), the dairy permeate powders covered by this standard may be modified in composition to meet the desired end-product composition, for instance, partial demineralization. However, compositional modifications beyond the minima or maxima specified above for lactose, nitrogen, milk fat, ash and moisture are not considered to be in compliance with the Section 4.3.3 of the *General Standard for the Use of Dairy Terms*.

4 FOOD ADDITIVES

4.1
The use of food additives is not permitted for dairy permeate powders covered by this standard.

4.2 Processing aids

The processing aids used in products covered by this standard shall comply with the *Guidelines on Substances used as Processing Aids* (CXG 75-2010).

5 CONTAMINANTS

The products covered by this Standard shall comply with the Maximum Levels for contaminants that are specified for the product in the *General Standard for Contaminants and Toxins in Food and Feed* (CXS 193-1995).

The milk used in the manufacture of the raw materials covered by this Standard shall comply with the Maximum Levels for contaminants and toxins specified for milk by the *General Standard for Contaminants and Toxins in Food and Feed* (CXS 193-1995) and with the maximum residue limits for veterinary drug residues and pesticides established for milk by the CAC.

6 HYGIENE

It is recommended that the product covered by the provisions of this standard be prepared and handled in accordance with the appropriate sections of the *General Principles of Food Hygiene* (CXC 1-1969), the *Code of Hygienic Practice for Milk and Milk Products* (CXC 57-2004) and other relevant Codex texts such as Codes of Hygienic Practice and Codes of Practice. The products should comply with any microbiological criteria established in accordance with the *Principles and Guidelines for the Establishment and Application of Microbiological Criteria Related to Foods* (CXG 21-1997).

7 LABELLING

In addition to the provisions of the *General Standard for the Labelling of Prepackaged Foods* (CXS 1- 1985) and the *General Standard for the Use of Dairy Terms* (CXS 206-1999) the following specific provisions apply.

7.1 Name of the food

The name of the food shall be dairy permeate powder. Products complying with the relevant descriptions in Section 2 and compositions in Section 3.3 may be named milk permeate powder and whey permeate powder, respectively.

Where appropriate in the country of sale, the name may be supplemented by the designation "lactose-rich deproteinized _____ powder", the blank being filled with the term dairy, whey or milk, as appropriate to the nature of the product.

7.2 Labelling of non-retail containers

Information required in Section 7 of this Standard and Sections 4.1 to 4.8 of the *General Standard for the Labelling of Prepackaged Foods* (CXS 1-1985), and, if necessary, storage instructions, shall be given either on the container or in accompanying documents, except that the name of the product, lot identification, and the name and address of the manufacturer or packer shall appear on the container. However, lot identification, and the name and address of the manufacturer or packer may be replaced by an identification mark, provided that such a mark is clearly identifiable with the accompanying documents.

8 METHODS OF SAMPLING AND ANALYSIS

For checking the compliance with this standard, the methods of analysis and sampling contained in the *Recommended Methods of Analysis and Sampling* (CXS 234-1999) relevant to the provisions in this standard, shall be used.